BIOLOGICAL PHYSICS SERIES

BIOLOGICAL PHYSICS SERIES

PUBLISHED VOLUMES

George J. Hademenos
Tarik F. Massoud

The Physics of Cerebrovascular Diseases

Biophysical Mechanisms of Development,
Diagnosis and Therapy

With Foreword by Fernando Viñuela

With 122 Illustrations, 6 in Color

Springer

George J. Hademenos
Department of Physics
University of Dallas
Irving, TX 75062
USA

Tarik F. Massoud
Department of Radiological Sciences
Endovascular Therapy Service
UCLA School of Medicine
Los Angeles, CA 90095-1721
USA

Hademenos, George J.
    Physics of cerebrovascular diseases : biophysical mechanisms of
development, diagnosis, and therapy / George J. Hademenos, Tarik F.
Massoud.
        p.   cm. — (International series in basic and applied
biological physics ; 1)
    Includes bibliographical references and index.
    ISBN 1-56396-558-5 (alk. paper)
    1. Cerebrovascular diseases.   2. Biophysics.   I. Massoud, Tarik F.
II. Title.   III. Series.
    [DNLM: 1. Cerebrovascular Disorders—physiopathology.
2. Cerebrovascular Disorders—therapy.   3. Cerebrovascular
Disorders—diagnosis.   4. Cerebrovascular Circulation—physiology.
5. Biophysics.   WL 355 H128p 1997]
RC388.5.H34   1977
618.8´1—dc21
DNLM/DLC
for Library of Congress   97-30653
                                                              CIP
Printed on acid-free paper.

© 1998 Springer-Verlag New York, Inc. AIP Press is an imprint of Springer-Verlag New York.

Production managed by Lesley Poliner; manufacturing supervised by Jacqui Ashri.
Typeset by Asco Trade Typesetting Ltd., Hong Kong from the author's Microsoft Word files.
Printed and bound by Maple-Vail Book Manufacturing Group, York, PA.
Printed in the United States of America.

9 8 7 6 5 4 3 2 1

ISBN 1-56396-558-5 Springer-Verlag New York Berlin Heidelberg   SPIN 10640080

To
Kelly and Alexandra, whose support I will always remember,
and
Dr. George Kattawar, whose advice I will never forget.

George J. Hademenos, Ph.D.

To
Susan and Zahra.

Tarik F. Massoud, M.D.

Foreword

It is a great pleasure and privilege to introduce this book to the scientific community interested in cerebrovascular diseases. Those of us working in interventional neuroradiology are accustomed to acquiring knowledge on an empirical basis. However, current research in cerebrovascular diseases is advancing beyond our usual methods and, to understand the complexity of these diseases, it is becoming increasingly necessary to understand the basic physical principles that govern most of the physiological and physiopathological phenomena that we face in our daily clinical practice. Drs. Hademenos and Massoud are two young, brilliant academicians who took on the challenge of depicting those essential biophysical mechanisms and presenting them in a comprehensive fashion to the neuroscience community.

The authors' goal in writing this book is to provide the scientist and the clinician with the fundamental physical principles of elasticity and hemodynamics as they are applied to the development, diagnosis, and modern therapy of complex cerebrovascular diseases, including cerebral aneurysms, arteriovenous malformations, and ischemic stroke. It is the first such attempt in this area, and the authors succeed admirably.

Dr. Hademenos and Dr. Massoud started a fruitful scientific association five years ago, in the Division of Interventional Neuroradiology at the UCLA School of Medicine. They achieved both national and international recognition with the development of mathematical and laboratory models to study the hemodynamic phenomena involved in aneurysms and arteriovenous malformations. They also incorporated therapeutic maneuvers to analyze postembolization anatomical and physiological outcomes and to predict technical or clinical complications such as aneurysm or AVM ruptures.

In doing so, Drs. Hademenos and Massoud overcame the numerous and difficult challenges in applying basic research, laboratory techniques, and clinical knowledge to the analysis of complex cardiovascular diseases.

Fernando Viñuela, M.D.
Professor and Director,
Division of Interventional Neuroradiology
UCLA School of Medicine and Medical Center

Series Preface

The field of biological physics is a broad, multidisciplinary, and dynamic one, touching on many areas of research in physics, biology, chemistry and medicine. New findings are published in a large number of publications within these disciplines, making it difficult for students and scientists working in biological physics to keep up with advances occurring in disciplines other than their own. The Biological Physics Series is intended therefore to be a comprehensive one covering a broad range of topics important to the study of biological physics. Its goal is to provide scientists and engineers with text books, monographs and reference books to address the growing need for information.

Books in the Biological Physics Series will emphasize frontier areas of science including molecular, membrane, and mathematical biophysics; photosynthetic energy harvesting and conversion; information processing; physical principles of genetics; sensory communications; automata networks, neural networks, and cellular automata. Equally important will be coverage of current and potential applied aspects of biological physics such as biomolecular electronic components and devices, biosensors, medicine, imaging, physical principles of renewable energy production, and environmental control and engineering.

We are fortunate to have a distinguished roster of consulting editors on the Editorial Board, reflecting the breadth of biological physics. We believe that the Biological Physics Series can help advance the knowledge in the field by providing a home for publications and that scientists and practitioners from many disciplines will find much to learn from the upcoming volumes.

Elias Greenbaum
Series Editor-in-Chief

Preface

Physics, in its native form, describes the science of "Mother Nature" or the elementary relations between actions and reactions of macroscopic and microscopic objects acted on by external forces. Although the discussion of physics presented in elementary courses involves a wide range of topics, corresponding examples presented to introduce and illustrate such concepts are limited to common phenomena observed on an everyday basis. Physics textbooks exist that describe the normal functions and processes of the human body; however, the importance of such knowledge becomes extremely useful to understand how and why human disease processes occur, factors involved in the development of such processes, and the rationale, invention, and implementation of therapies to effectively treat the disease, improving both patient safety and outcome.

According to the most recent statistics, cerebrovascular diseases are the third leading cause of death, ranking behind heart attack and all forms of cancer.[1] Cerebrovascular diseases adversely afflict the blood vessels and blood flow within the brain. The neurological consequences of cerebrovascular diseases are devastating, if not fatal, and significantly impact the cost and overall status of health care. Current knowledge of cerebrovascular disease has advanced due to theoretical, experimental, and clinical observations that have led to substantial strides in understanding the pathogenesis of such diseases, identification of risk factors, improvements in diagnostic and therapeutic techniques, and reduction of morbidity and mortality rates. All of these aspects have evolved primarily as a result of a thorough and in-depth understanding of the biophysical interactions and phenomena associated with cerebrovascular diseases.

This textbook serves as an introductory review of physics as it pertains to cerebrovascular diseases. Although there are a variety of dedicated textbooks that sufficiently address clinical observations and various physical concepts such as elasticity and hemodynamics, no textbook exists presently which addresses these fundamental physical concepts as they are applied directly to the development, management, diagnosis and therapy of cerebrovascular diseases. This textbook is intended to fill this void between clinical or experimental observations and their scientific origins, serving to benefit both the scientist and clinician. More specifically, this book is designed to provide the reader (both scientist and clinician) a general scientific foundation for all aspects of cerebrovascular diseases including: (1) the physical basis and mechanisms for the development and clinical consequences of cerebrovascular diseases; (2) a description of current and future therapies for cerebrovascular diseases; (3) numerical, experimental, and clinical techniques employed to investigate cerebrovascular diseases; (4) a review of blood flow dynamics, methods of measurement, and its role in cerebrovascular diseases; and, (5) a survey of current imaging methods for the diagnosis and management of cerebrovascular diseases. The scientists, particularly those interested in unique applications of science or a research career in medicine, will gain: (1) an introductory understanding of the role of physics in these diseases and (2) a basis for theoretical and experimental research geared toward advancing the knowledge of cerebrovascular diseases. The clinician as well as the medical student/intern will find this text an enlightening yet important supplement to their clinical education and patient management by acquiring a different perspective in the development, diagnosis, and treatment of cerebrovascular diseases.

This book was written as a one- or two-semester textbook intended for the upper level undergraduate/entry level graduate student majoring in physics, biophysics, biomedical engineering, medical physics, biomechanics, biology, physiology, pathology, or pre-medicine. This book would also be of benefit to medical students/interns and nursing students. This book is not intended specifically for medical specialists in neurology, cardiology, neuroradiology, cardiovascular surgery, and neurosurgery; nevertheless, it should serve them as a useful overview of the physical aspects of cerebrovascular diseases. Although it is assumed that the reader has had at least minimum exposure to the biological and physical sciences, the student is not required to have formal coursework or specialized training, as background and introductory material is addressed in preliminary chapters. This book addresses only the elementary aspects of the development, diagnosis, and therapy of cerebrovascular diseases, i.e., physical/biophysical mechanisms. This book is *not* a clinical authoritative review of these diseases and is not intended to replace, but instead is to supplement, current textbooks used in the educational curriculum of clinical specialists in the neurosciences. It should also be mentioned that the coverage of topics in this book is by no means exhaustive and many other aspects in the study of cerebrovascular diseases exist but were not covered due to the size and scope of the book. Although mentioned briefly, detailed analysis of the molecular biology, pharmacology, pathology, etc. is not included. Interested readers are encouraged to consult corresponding reference citations involving these fields for further information.

The textbook is structured in two distinct parts: (1) an overview of the fundamental mathematics and physics required to describe and characterize cerebrovascular diseases, and (2) an application of physics as it pertains specifically to aspects of cerebrovascular diseases. The first part of this book, consisting of Chaps. 1–4, provides a concise introduction and review of the elementary mathematical fundamentals that will be used in subsequent chapters (Chap. 1), the physics of elasticity (Chap. 2), circulatory physiology (Chap. 3), and the physics of blood flow or hemodynamics (Chap. 4). In the second part of this book, represented by Chaps. 5–7, chapters are dedicated to the following cerebrovascular diseases: stroke (Chap. 5), aneurysms (Chap. 6), and arteriovenous malformations (Chap. 7). These topics are discussed here to provide a clear and concise overview of the material in a manner designed to demonstrate the physical concepts and mechanisms involved in cerebrovascular diseases. It should be stated that the topics covered in this book can be extremely detailed, complex, and, in most instances, require textbooks of their own to adequately describe the material. Both the size and scope of this book prevent the inclusion of expanded coverage on these topics. In response to this, every attempt is made to guide the student or reader to more comprehensive standard textbooks on subjects addressed briefly in this book.

A Solutions Manual for this book is available free of charge. To request it, please write to Maria Taylor, Springer-Verlag, 175 5th Ave., New York, NY, 10010.

George J. Hademenos, Ph.D.
Department of Physics
University of Dallas

Tarik F. Massoud, M.D.
Endovascular Therapy Service
Department of Radiological
    Sciences
UCLA School of Medicine

Reference

1. *1996 Heart and Stroke Facts* (American Heart Association, Dallas, 1996).

Acknowledgments

The authors would like to express their sincere appreciation to Suzie El-Saden, M.D. for her assistance and technical expertise in the review of portions of the manuscript and Lynne Olson for the graphical illustrations.

Contents

1
Mathematics Fundamentals

1.1 Introduction

As a scientific discipline, the laws of physics are expressed typically in terms of qualitative reasoning and observations and rely on mathematics to translate these laws into quantitative values. Mathematics allows the student to grasp intuitively and intellectually the physical concepts and understand the effects of various factors on the particular physics problem. An example of such a physics problem encountered commonly in daily activities is projectile motion. Sports enthusiasts who are familiar with physics even on an introductory level know that the path of a ball (golf ball, football, basketball, baseball, etc) or projectile thrown or forced in motion is parabolic, in nature, and attains a maximum range for a projection angle of 45°. The parabolic path of the projectile can be expressed mathematically in equation form as[1]

$$y = (\tan \theta_0)x - \frac{g}{2v_0^2 \cos^2 \theta_0} x^2,$$

where x and y are the coordinates of the particle's position, θ_0 is the projection angle, v_0 is the initial velocity, and g is the acceleration of an object due to gravity. From this relation, one could obtain specific information concerning the flight of the projectile, such as the range obtained by the projectile given by

$$R = \frac{v_0^2}{g} \sin 2\theta_0,$$

which is maximum when $\sin 2\theta_0 = 1$, which corresponds to $2\theta_0 = 90°$ or a launching angle θ_0 of 45°.[1]

The level of complexity of mathematics increases correspondingly with that of the physics. Returning to our example of projectile motion, if a wind is present that provides a resistant force against the projectile or if the projectile is rotating, the problem of determining the maximum height, maximum range, or time to complete a projectile path becomes more complicated with regard to its physics and mathematics. As long as the mathematics can be expressed adequately to represent the physical behavior and corresponding factors included in such a problem, the quantitative solution to any aspect of the involved physics is readily available.

This chapter provides a brief description of the mathematics that will be used to describe and exemplify the physics of cerebrovascular diseases and related concepts covered in this textbook. The field of mathematics consists of a variety of topics that, in themselves, cannot be fully appreciated by the brief attention afforded to them in this chapter and are presented, in most cases, as entire books dedicated solely to that topic. Since this chapter is entitled "Mathematics Fundamentals," attempts will be made to address briefly a specific mathematical aspect or topic that will be instrumental in grasping a working knowledge and understanding of the biophysics of these disease processes. The purpose of this chapter is not to provide detailed instruction of the mathematical topics but to demonstrate and illustrate applications of these topics to the physics encountered in later chapters. As the individual mathematics topics are explained in the subsequent sections, an effort will be made to (1) describe the mathematical topics in a clear and concise fashion with applications and examples to the physical sciences; (2) identify later sections in the textbook where this form of mathematics will be applicable; and (3) reference relevant textbooks to provide the reader with a more detailed treatment and explanation of the mathematical topic. More specifically, the mathematical topics that will be addressed in this chapter include functional analysis, differential and integral calculus, ordinary and partial differential equations, vector analysis, matrices, Fourier analysis, and complex variables.

1.2   Functional Analysis

In studying physical concepts, we become interested in both the qualitative and quantitative influence of one variable on the behavior and stability of a given system. Thus, the purpose and overall objective of such investigations is the proper characterization of the system through the derivation of functional relationships among all related variables. A function represents a mathematical relationship between two variables, e.g., x and y, and is denoted typically as $y = f(x)$. Suppose, for example, we wanted to determine the influence of tube radius R and tube length L on the rate of fluid flow, Q, through the tube. Although there exist variables such as fluid density and viscosity that are embedded in the system function as constants, theory and experimental data reveal a relationship among the two primary variables R and L and the system variable Q according to the equation

$$Q = \frac{\pi \Delta P R^4}{8L\eta},$$

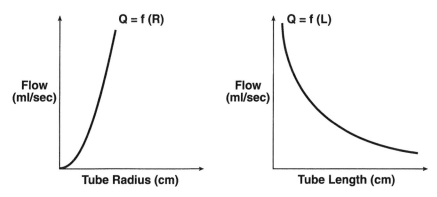

FIGURE 1.1. Two-dimensional graphical plots of the function for fluid flow in terms of the parameters (A) vessel radius R and (B) vessel length L. With respect to the radius, fluid flow exhibits an R^4 dependence with respect to vessel radius and a $1/L$ dependence with respect to vessel length.

where Q is the volumetric flow rate, ΔP is the pressure gradient, and η is the fluid viscosity. The relationship, known as Poiseuille's law, will be covered in more detail in Chap. 4. Let us assume that, since we are concerned primarily with cerebrovascular physiology in this book, the fluid is blood with a viscosity $\eta = 3.5$ centipoise (cp). The rate of fluid flow, Q, is a dependent variable since its value depends primarily upon the values of R and L, which are independent variables and can be expressed as a three-dimensional function,

$$Q = f(R, L)$$

but can also be expressed as a function of either of the two variables, either $Q = f(R)$ or $Q = f(L)$, while holding the opposing variable constant. When displayed graphically on a rectangular coordinate system with the two axes representing the two variables Q and either R or L, the functions, defined by either $Q = f(R)$ or $Q = f(L)$ are described by continuous curves in two dimensions, as shown in Fig. 1.1. This example illustrates the strong dependence of flow on the tube radius. The importance of this relation will become evident in subsequent chapters.

Consider the function describing a line or

$$y = mx + b.$$

Two important features of a function illustrated by the above equation are the y intercept (b) and the slope of the function or curve (m). Ideal functions originate at the origin (0,0) of the coordinate system. Most functions, however, originate at a point along the y axis other than 0. The value along the y axis, known as the y intercept, is given by the parameter b. The slope defines the rate of change of the curve in the y axis with respect to the corresponding change in the x direction and is given mathematically as

$$m = \frac{\Delta y}{\Delta x}.$$

The slope of a mathematical function depends on the type of function, i.e., polynomial function, power function, trigonometric function, etc. Poiseuille's law of fluid flow is a power function, given by $y = ax^n$ where a is a constant and n is any real number. The slope of a power function can be calculated by taking the logarithms of both sides of the function:

$$\log y = n \log x + \log a$$

where n now becomes the slope:

$$n = \frac{\Delta(\log y)}{\Delta(\log x)}$$

This is, in essence, the basis for the derivative and differential calculus that will be explained in the next section. Also, one can obtain the same information concerning the slope of a function from a three-dimensional (3D) function using partial differentiation, which will be discussed in Sec. 1.6.

As stated previously, inclusion of both parameters R and L represents Q as a 3D function in space where the function is now defined as $z = f(x,y)$ and the curve now becomes a surface. Almost as important as the function itself are the geometric boundaries through which the function is defined. This can be understood by the following illustrative example. A point in space bounded by a rectangular coordinate system is given as $x = x$, $y = y$, and $z = z$. However, if that same point were now defined within a different geometry such as a cylinder or sphere, it becomes a cumbersome, although possible, task to define the point in rectangular coordinates. It thus becomes a matter of convenience to denote any point within a sphere or cylinder in terms of the variables that define its geometry in space through coordinate transformations among different geometries. Given the geometry of the cylinder and sphere depicted in Fig. 1.2, the three-dimensional coordinates of a point in space that were defined originally as x, y, and z, in the rectangular coordinate system are now redefined for the cylinder in cylindrical coordinates (r, ϕ, z) as

$$x = r\cos\phi, \quad y = r\sin\phi, \quad z = z,$$

where $r \geq 0$, $0 \leq \phi < 2\pi$, $-\infty < z < \infty$, and for the sphere in spherical coordinates (r, θ, ϕ) as

$$x = r\sin\theta\cos\phi, \quad y = r\sin\theta\sin\phi, \quad z = r\cos\theta,$$

where $r \geq 0$, $0 \leq \theta \leq \pi, 0 \leq \phi < 2\pi$. This importance and applications of coordinate transformations will become evident with approximations of cylindrical geometry for blood vessels (Chaps. 2 and 4), spherical geometry

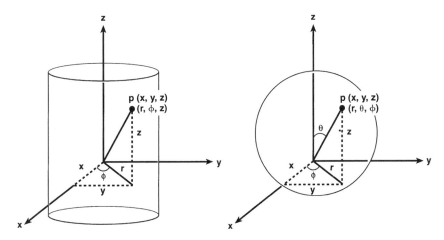

FIGURE 1.2. Schematic diagram depicting the three-dimensional coordinates of a point P in a (A) cylindrical coordinate system and (B) spherical coordinate system.

for aneurysms (Chap. 6), and the derivation of Laplace's law for elastic spherical and cylindrical geometries (Chap. 2).

Although numerous possibilities satisfy the criteria for a function, there are five general classifications of functions and are listed below.

1.2.1 Polynomial Functions

Polynomial functions are the most general of the five types of functions and are of the form

$$f(x) = a_0 + a_1 x + a_2 x^2 + a_3 x^3 + \cdots + a_n x^n,$$

where a_n are coefficients and n is an integer that describes the degree or order of the polynomial. The most common polynomial function is

$$f(x) = a_0 + a_1 x$$

Individual terms of higher order polynomial functions are known as power functions defined by:

$$f(x) = ax^n$$

where a is a constant and n is a real number, representing the mathematical power of the function. Figure 1.3(A) displays graphically the more common polynomial functions. Polynomial and power functions represent a wide range of mathematical functions and can be found in numerous physical applications as they pertain to cerebrovascular diseases including elasticity (Chap. 2) and hemodynamics (Chap. 4). One example, in particular, was *Poiseuille's law* described previously for fluid flow through a tube.

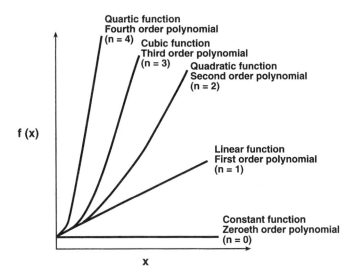

FIGURE 1.3. Illustrative examples of the five general types of mathematical functions. (A) The first five orders of polynomial functions: $n = 0$: constant; $n = 1$: linear; $n = 2$: quadratic; $n = 3$: cubic; $n = 4$: quartic. (B) Trigonometric functions: $\sin(x), \cos(x), \tan(x), \cot(x), \sec(x), \csc(x)$. (C) Exponential function: Ca^{bx}. (D) Logarithmic function: $\log_a x$. (E) Special function: Bessel function: $J_1(x)$. This solution to Bessel's differential equation is of the first kind and, in general, of order n. In this graph, Bessel functions of the first kind of order 1 are displayed.

1.2.2 Trigonometric Functions

With the periodic nature of cardiac physiology and blood flow, one can deduce easily the potential importance of trigonometric functions and their role in cerebrovascular diseases. Trigonometric functions are employed typically in the description of periodic or oscillatory phenomena. The family of trigonometric functions include

$$f(x) = \sin(x), \ \cos(x), \ \tan(x), \ \cot(x), \ \sec(x), \ \csc(x),$$

where the argument x, is a given angle. Figure 1.3B provides a graphic example of $\sin x$, $\cos x$, $\tan x$, $\cot x$, $\sec x$, $\csc x$ illustrating the periodic nature of trigonometric functions.

Some of the more important relations between the trigonometric functions are

$$\sin(-x) = -\sin x, \quad \cos(-x) = \cos x, \quad \tan(-x) = -\tan x,$$

$$\tan x = \frac{\sin x}{\cos x}, \quad \sin^2 x + \cos^2 x = 1, \quad \sec^2 x - \tan^2 x = 1,$$

$$\sin(x \pm y) = \sin x \cos y \pm \cos x \sin y, \quad \cos(x \pm y) = \cos x \cos y \mp \sin x \sin y.$$

$$\textbf{sin } \theta = \frac{a}{c} = \frac{\text{opposite}}{\text{hypotenuse}}$$

y = sin x

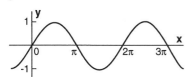

$$\textbf{cos } \theta = \frac{b}{c} = \frac{\text{adjacent}}{\text{hypotenuse}}$$

y = cos x

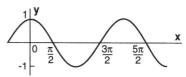

$$\textbf{tan } \theta = \frac{a}{b} = \frac{\text{opposite}}{\text{adjacent}}$$

y = tan x

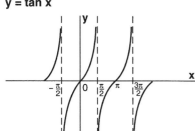

$$\textbf{cot } \theta = \frac{b}{a} = \frac{\text{adjacent}}{\text{opposite}}$$

y = cot x

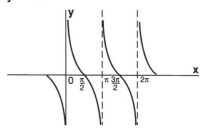

$$\textbf{sec } \theta = \frac{c}{b} = \frac{\text{hypotenuse}}{\text{adjacent}}$$

y = sec x

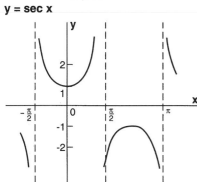

$$\textbf{csc } \theta = \frac{c}{a} = \frac{\text{hypotenuse}}{\text{opposite}}$$

y = csc x

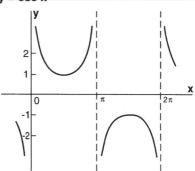

FIGURE 1.3 (*continued*)

Other forms of trigonometric functions such as inverse trigonometric functions ($\sin^{-1}x$, $\cos^{-1}x$, $\tan^{-1}x$, $\cot^{-1}x$, $\sec^{-1}x$, and $\csc^{-1}x$) and hyperbolic functions ($\sinh x$, $\cosh x$, $\tanh x$, $\coth x$, $\mathrm{sech}\, x$, $\mathrm{csch}\, x$) have minimal applications to the topic at hand and will be introduced only as they are presented.

Trigonometric functions (together with exponential functions described

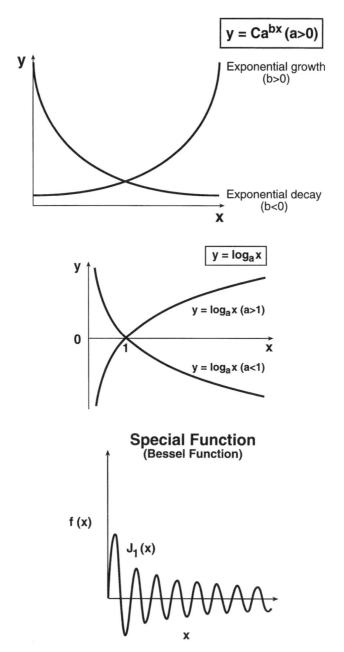

FIGURE 1.3 (*continued*)

below) serve a significant role in the solution of differential equations and the corresponding description of physiological processes including the propagation of waves through elastic media (Chap. 2) and pulsatile blood flow (Chap. 4).

1.2.3 Exponential Functions

Exponential functions are defined mathematically as $f(x) = Ca^{bx} (a > 0)$, where a, b, and C are constants. Exponential functions describe many biological processes, particularly when $b > 0$ (growth processes) and when $b < 0$ (decay processes). They are especially useful when a is equal to $e = 2.718$. In this case, exponential functions are expressed mathematically as $f(x) = Ce^{bx}$. In Fig. 1.3C, exponential functions representing growth and decay processes are displayed. Although they are not periodic functions, exponential functions can be expressed in terms of trigonometric functions according to

$$e^{ix} = \cos x + i \sin x,$$

where i is a complex number $(= \sqrt{-1})$. Complex numbers and their application to the description of physical systems will be described later in this chapter. Thus, exponential functions become just as important as trigonometric functions as the basis for solutions of differential equations. Many physical problems such as radioactive decay, commonly seen in nuclear medicine imaging techniques, and electrical circuits (Chap. 4) involve exponential functions in their solutions.

1.2.4 Logarithmic Functions

Logarithmic functions are related to the previous type of function in that they are the inverse of exponential functions. Related to the exponential function $f(x) = e^x$, the logarithm of $f(x) = \log_e x$ and can be simplified further to $f(x) = \ln x$, shown in Fig. 1.3D. The physical applications of logarithmic functions as they pertain to this book will be described later in this chapter and in the description of biomathematical models for arteriovenous malformations (Chap. 7).

1.2.5 Special Functions

Special functions are mathematical functions, derived typically from the solution of differential equations, that are applicable only to unique sets of problems and are not as used widely as polynomial or trigonometric functions. An excellent example of special functions as applicable to the topics presented in this book is that of Bessel's function. As one can see from the graph presented in Fig. 1.3E, Bessel's function exhibits a damped oscillatory behavior that would serve a potential purpose in characterizing blood flow.

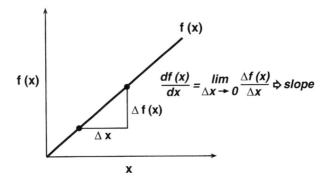

FIGURE 1.4. Graphical description of the derivative of a mathematical function.

Bessel's function is found in solutions to problems involving pulsatile fluid dynamics and will be described in more detail in Chap. 4. For further details on the various types of special functions and their applications to the physical sciences, the reader is directed to textbooks entirely devoted to this topic.[2-5]

1.3   Differential Calculus

The basis for differential calculus is the derivative, an entity that is fundamental to the characterization of dynamic physical processes. The derivative of a function, depicted symbolically as $df(x)/dx$ ($= dy/dx$) or $f'(x)$, yields the tangent of a mathematical function or the rate of increase (or decrease) along the y axis with respect to the x axis (Fig. 1.4). In the above expression, dy/dx are termed *differentials*. Although the differentials dx and dy are the same as Δx and Δy in meaning, they represent infinitesimal changes along their respective axes. The derivative dy/dx can be equated with the slope $\Delta y/\Delta x$ only in the limit as Δx becomes very small and approaches 0. This is expressed typically as

$$\frac{dy}{dx} = \lim_{\Delta x \to 0} \frac{\Delta y}{\Delta x}.$$

Upon closer inspection, this, in essence, is the definition of a slope, and the two are, in fact, one and the same. A tabulated list of the formulas for derivatives for the most common functions described in Sec. 1.2 is displayed in Table 1.1.

A physical application of the derivative can be illustrated with the problem of motion of a particle or object. Consider an object that moves in space

TABLE 1.1. Differentiation formulas for common mathematical functions.

Polynomial functions

$$\frac{d}{dx}(a_0 x^n) = na_0 x^{n-1}$$

$$\frac{d}{dx}(a_0 u^n) = na_0 u^{n-1}\frac{du}{dx}$$

Trigonometric functions

$$\frac{d}{dx}(\sin u) = \cos u\,\frac{du}{dx}$$

$$\frac{d}{dx}(\cos u) = -\sin u\,\frac{du}{dx}$$

$$\frac{d}{dx}(\tan u) = \sec^2 u\,\frac{du}{dx}$$

$$\frac{d}{dx}(\cot u) = -\csc^2 u\,\frac{du}{dx}$$

$$\frac{d}{dx}(\sec u) = -\sec u \tan u\,\frac{du}{dx}$$

$$\frac{d}{dx}(\csc u) = -\csc u \cot u\,\frac{du}{dx}$$

Exponential functions

$$\frac{d}{dx}(a^u) = a^u \ln a$$

$$\frac{d}{dx}(e^u) = e^u\,\frac{du}{dx}$$

Logarithmic functions

$$\frac{d}{dx}(\log u) = \frac{\log_a e}{u}\frac{du}{dx}, \quad a \neq 0, 1$$

$$\frac{d}{dx}(\ln u) = \frac{1}{u}\frac{du}{dx}$$

---

according to the function

$$|\mathbf{x}| = f(t) = 4t^3 + 10t^2 + 6.$$

The boldface type implies that the parameter represents a vector quantity (Sec. 1.7). Referring to Table 1.1, the derivative of a polynomial function is given as

$$\frac{df(x)}{dx} = \frac{d(x^n)}{dx} = nx^{n-1}.$$

Then the velocity of the object is the derivative of $f(t)$ or $\mathbf{x}$,

$$|\mathbf{v}| = \frac{d\mathbf{x}}{dt} = 12t^2 + 20t,$$

and the acceleration of the object is the derivative of the velocity $\mathbf{v}$, the second derivative of $\mathbf{x}$, or

$$|\mathbf{a}| = \frac{d\mathbf{v}}{dt} = \frac{d^2\mathbf{x}}{dt^2} = 24t + 20,$$

where the superscript 2 refers to the order of the derivative. The applications of a derivative include identification of maximum and minimum points within a function, description of the rate of change of a function $f(x)$ with respect to x, and determination of the velocity and acceleration of a particle given its position, as just demonstrated by the above example.

1.4   Integral Calculus

The integral is to integral calculus as the derivative is to differential calculus and serves just as important a role in the characterization of dynamic processes. Integration of a given function $f(x)$, in essence, is the inverse of differentiation and is defined mathematically by

$$\int f(x)\, dx,$$

where dx is the differential along the x axis. The integral represents the area under the curve $y = f(x)$ bounded by limits of x_1 and x_2 along the x axis and predefined limits along the y axis. The area under the curve is subdivided into equal regions of size dx, as shown in Fig. 1.5. By integrating the function, one is, in effect, summing all of the dx regions bounded by the function curve.

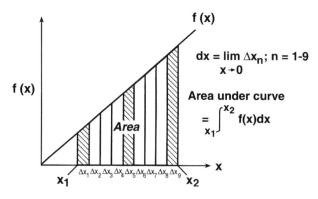

FIGURE 1.5. Graphical description of the integral of a mathematical function.

TABLE 1.2. Integration formulas for
common mathematical functions.

Polynomial functions

$$\int u^p\, du = \frac{u^{p+1}}{p+1}, \quad p \neq 1$$

$$\int u^{-1}\, du = \int \frac{du}{u} = \ln u$$

Trigonometric functions

$$\int \sin u\, du = -\cos u$$

$$\int \cos u\, du = \sin u$$

$$\int \tan u\, du = -\ln \cos u$$

$$\int \cot u\, du = \ln \sin u$$

$$\int \sec u\, du = \ln(\sec u + \tan u)$$

$$\int \csc u\, du = \ln(\csc u - \cot u)$$

Exponential functions

$$\int a^u\, du = \frac{a^u}{\ln a}, \quad a > 0, \quad a \neq 1$$

$$\int e^u\, du = e^u$$

Logarithmic functions

$$\int \ln x\, dx = x \ln x - x$$

To show the relationship between the integral and derivative, we shall use the same illustrative example of particle motion presented in Sec. 1.3. However, to demonstrate integration, we must begin in a manner opposite to that presented earlier. Thus, the equation for the acceleration of the particle,

$$|\mathbf{a}| = \frac{d\mathbf{v}}{dt} = \frac{d^2\mathbf{x}}{dt} = 24t + 20,$$

is integrated to obtain the velocity of the particle. A tabulated list of the integral formulas for the most common mathematical functions are given in Table 1.2. The integral for a polynomial, given by

$$\int u^p\, du = \frac{u^{p+1}}{p+1}, \quad p \neq -1$$

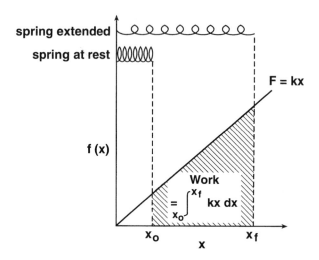

FIGURE 1.6. A physical application of an integral for the calculation of the work done by stretching a spring.

is applied to the particle acceleration and, upon solving the integral, yields the velocity

$$|\mathbf{v}| = \frac{d\mathbf{x}}{dt} = 12t^2 + 20t.$$

Integrating once again using the integral for a polynomial function gives the position of the particle:

$$|\mathbf{x}| = f(t) = 4t^3 + 10t^2 + C,$$

where C is an integration constant $(= 6)$.

The physical significance of the integral can also be illustrated by considering the example of Hooke's law. Hooke's law describes the force F required to stretch a spring a distance x and is given by

$$F = -kx,$$

where k is the force constant of the spring. Figure 1.6 shows the graph of force versus stretch distance. The negative sign corresponds to the opposite direction with which the restoring force of the spring acts against the external stretching force. The area under the curve presented in Fig. 1.6 represents the work done by the external stretching force and can be determined quantitatively by integrating Hooke's law over a given region from $x = x_0$ to $x = x_f$:

$$W = \tfrac{1}{2}kx_f^2 - \tfrac{1}{2}kx_0^2.$$

The minus sign was omitted from the integral by reversal of the direction along which the work was performed. Assuming that x_0 represents the origin $(x = 0)$, then the work done by the spring is

$$W = \tfrac{1}{2}kx^2.$$

Extending this concept, suppose one is only interested in obtaining the work (or area under the curve of Hooke's law) over a defined region, say from $x = 0$ to $x = 4$. In this case, the work is determined in a similar fashion with the substitution of the numerical limits in the evaluation of the integral:

$$W = \tfrac{1}{2}k(4)^2 - \tfrac{1}{2}k(0)^2 = 8k.$$

1.5 Ordinary Differential Equations

As stated previously, a variety of physical phenomena are dynamic in nature and thus exhibit a time-dependent behavior that can be described using ordinary differential equations. The term *ordinary* is used to delineate between partial differential equations (discussed in the next section). From this point onward, the term ordinary will be dropped from the topic description. Differential equations are defined simply as mathematical equations that contain derivatives. The solution of a differential equation should be such that the derivative of a function is either identical or maintains a distinct similarity to the original function. Generally speaking, the two types of mathematical functions that satisfy this criteria in most cases are exponential functions and trigonometric functions. Several illustrative examples of applications of differential equations to physical phenomena include radioactive decay, electrical circuits, and harmonic motion.

1.5.1 Radioactive Decay

Radioactive decay is a nuclear phenomenon exhibited by radioisotopes or elements with an atomic number Z greater than that of lead $(Z = 82)$. Radioactive elements contain generally more Z than mass number A implying an instability in internal energy. The radioactive element, in nature, strives toward a stabilized state of existence and, in the process, spontaneously emits particles (photons and charged and uncharged particles) in the transformation to a different nucleus and hence a different element. This process is referred to as radioactive decay and is dependent on the amount and identity of the radioactive element. Assume we have a sample of a radioactive element where the following variables are identified: N, the number of radioactive atoms in the sample, and, λ, the decay constant or constant describing the rate of decay unique to the radioactive element. Given this, the rate of change of the number of radioactive atoms (dN/dt) is equal to the rate of decay or decay constant multiplied by the number of

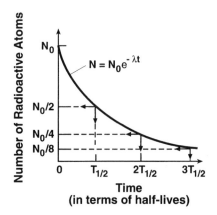

FIGURE 1.7. Graph depicting the decay of a typical radioactive element with respect to its half-life.

radioactive atoms, as shown in Fig. 1.7, written in equation form as

$$\frac{dN}{dt} = -\lambda N.$$

Solution of this equation can be accomplished by first manipulating like terms to the same side of the equation:

$$\frac{dN}{N} = -\lambda t.$$

Integrating the above equation over the limits of N and N_0 yields

$$\ln\left(\frac{N}{N_0}\right) = -\lambda t.$$

This equation can be simplified by taking the exponential of both sides:

$$\frac{N}{N_0} = e^{-\lambda t},$$

where N_0 is the original population of radioactive atoms. Further simplification of the equation results in the familiar equation of radioactivity:

$$N = N_0 e^{-\lambda t}.$$

As can be seen, the solution to the radioactive decay equation contains an exponential function.

1.5.2 Electrical Circuits

An electrical circuit is expressed in equation form according to the physical interactions of the charge q (from the flowing current i) with the component present in the electrical circuit. Consider, for example, an electrical circuit

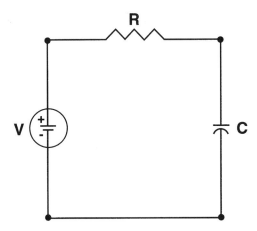

FIGURE 1.8. Schematic diagram of an electrical circuit with resistance (R), capacitance (C), and voltage (V) elements.

with a resistance R and a capacitor C, powered by a voltage source V, shown in Fig. 1.8. Analysis of an electrical circuit is typically performed according to Kirchoff's two laws:

Junction rule: The sum of all currents entering a junction or node must be equal to the sum of all currents exiting the node.

Loop rule: The algebraic sum of potential drops encountered while traversing the circuit in a clockwise direction must be equal to zero.

Application of these rules to the electrical circuit in Fig. 1.8 results in an equation by which all parameters can be characterized[6]

$$V = iR + \frac{q}{C}.$$

Since the current is related to the charge by

$$i = \frac{dq}{dt} = \frac{V}{R} e^{-t/RC},$$

the differential equation can be rewritten in terms of charge according to

$$V = R\frac{dq}{dt} + \frac{q}{C}.$$

The solution for charge of the above equation yields[1]

$$Q = CV(1 - e^{-t/RC})$$

and can also be expressed in terms of the current i by taking its derivative,

$$i = \frac{dq}{dt} + \frac{V}{R} e^{-t/RC}.$$

Again, an exponential function appears in the equation from the solution of the electrical circuit.

1.5.3 Harmonic Oscillator

A harmonic oscillator describes any physical system capable of producing oscillatory or periodic phenomena such as a mass fastened to a spring. In this system, the mass is at rest and the system is in equilibrium. When pulled a finite distance x past the point of equilibrium and released, the object responds with oscillatory or periodic motion about the point of equilibrium until the energy of the spring ultimately dissipates, subsequently stopping the imposed motion of the mass, as shown in Fig. 1.9. We have previously defined Hooke's law as

$$F = -kx.$$

The force applied to the attached mass is defined, according to Newton's second law, by the product of its mass and its acceleration, which, in turn, can be expressed as the second derivative of position versus time:

$$F = m\frac{d^2x}{dt^2}.$$

Substituting into the above relation for Hooke's law reveals

$$m\frac{d^2x}{dt^2} + kx = 0$$

with the solution to this equation given by

$$x = A\sin\omega_0 t + B\cos\omega_0 t,$$

where ω_0 is the angular frequency of the oscillating mass. This is an example of a differential equation solved in terms of trigonometric functions.

A number of different mathematical techniques and approaches, coupled with the implementation of appropriate boundary conditions, exist to solve differential equations. These solution methods can be cumbersome mathe-

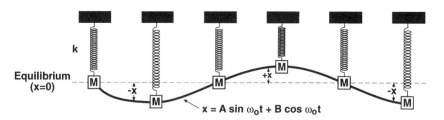

FIGURE 1.9. Illustrative example of the dynamics of a harmonic oscillator. When set in motion, the mass of the harmonic oscillator oscillates in a periodic nature that can be described by a trigonometric function.

matically, prone to computational errors, and time consuming. When confronted with a differential equation requiring such mathematical operations, the first approach toward a solution should be the implementation of educated guesses in the form of trigonometric or exponential expressions into the equation to see if the equation holds true. Should this be unsuccessful, another possible approach to the solution of differential equations involves the consultation of appropriate mathematical handbooks. Other approaches toward the solution include the use of computer software packages such as *MATHEMATICA* (Wolfram Research Inc., Champaign, IL) or *MAPLE V* (Waterloo Maple Software, Pacific Grove, CA), which can numerically solve such equations and also provide solutions in closed form. Additional information on differential equations can be found in a wide assortment of textbooks.[7-9]

1.6   Vector Analysis

Although it is recognized as a formal topic of mathematics, vector analysis also plays a fundamental role in physics. Vectors are measurements of physical quantities, processes, and interactions that exhibit both magnitude and direction of a displacement. Several examples of vector quantities include velocity, acceleration, and force. This is in contrast to scalar quantities or physical parameters that exhibit only magnitude. Examples of scalar quantities include distance, speed, and time. A more complete list of examples of vector and scalar quantities encountered commonly in physical applications is given in Table 1.3.

Vectors represent an alternative of describing a function and are expressed typically in a three-dimensional coordinate system as

$$\mathbf{V} = V_x \mathbf{i} + V_y \mathbf{j} + V_z \mathbf{k},$$

where V_x, V_y, and V_z are coefficients representing vector measurements and $\mathbf{i}$, $\mathbf{j}$, and $\mathbf{k}$ are unit vectors along the x, y, and z coordinates, as shown in Fig. 1.10. The magnitude of a vector is the square root of the sum of the square of its components, or

$$|\mathbf{V}| = \sqrt{(V_x^2) + (V_y^2) + (V_z^2)}.$$

Coordinate transformations can be applied to a vector as follows.
   Cylindrical coordinates:

$$V_r = \mathbf{i} \cos \phi + \mathbf{j} \sin \phi,$$

$$V_\phi = -\mathbf{i} \sin \phi + \mathbf{j} \cos \phi,$$

$$V_z = \mathbf{k}.$$

TABLE 1.3. Scalar and vector quantities of selected physical parameters.

Scalar quantities	Vector quantities
Measurement	
Time	
Mass	
Area	
Volume	
Kinematics	
Distance	Displacement
Speed	Velocity
	Acceleration
Dynamics	
Energy	Force
Work	Momentum
Moment of inertia	Torque
Frequency	Electric current
Viscosity	Electrostatic field
	Electromagnetic field

Spherical coordinates:

$$V_r = \mathbf{i} \sin\theta \cos\phi + \mathbf{j} \sin\theta \sin\phi + \mathbf{k} \cos\theta,$$

$$V_\theta = \mathbf{i} \cos\theta \cos\phi + \mathbf{j} \cos\theta \sin\phi - \mathbf{k} \sin\theta,$$

$$V_\phi = -\mathbf{i} \sin\phi + \mathbf{j} \cos\phi.$$

Since vectors are functions, all mathematical topics discussed previously, such as differential calculus, integral calculus, and partial differentiation are readily applicable to vectors, as shown by the following.

Vector differentiation:

$$\frac{d\mathbf{V}}{dt} = \frac{dV_x}{dt} \mathbf{i} + \frac{dV_y}{dt} \mathbf{j} + \frac{dV_z}{dt} \mathbf{k},$$

$$\frac{\partial \mathbf{V}}{\partial t} = \frac{\partial V_x}{\partial t} \mathbf{i} + \frac{\partial V_y}{\partial t} \mathbf{j} + \frac{\partial V_z}{\partial t} \mathbf{k}.$$

Vector integration:

$$\int \mathbf{V}\, dt = \mathbf{i} \int V_x\, dt + \mathbf{j} \int V_y\, dt + \mathbf{k} \int V_z\, dt.$$

The introduction of vectors in this chapter has revolved about the previously described topics to place everything in perspective, which is necessary for a brief synopsis of a vast amount of information. However, one is

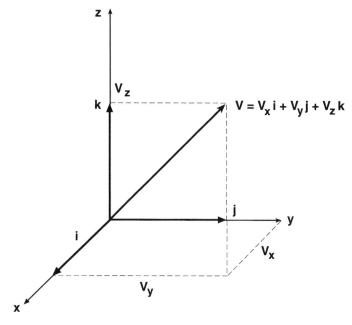

FIGURE 1.10. Graph showing the magnitude, direction, and individual components of a vector.

confronted with the fact that, when dealing with problems of a physical nature, the presence of several factors must be accounted for and implemented into subsequent calculations. In other words, if a problem requires the use of several vectors, how can these vectors be resolved and added to reveal meaningful results?

1.6.1   Resolution of Vectors

The vector is represented graphically by an arrow with the tail situated at the origin, pointing in a direction dictated by the problem with the size or length of the vector corresponding to its magnitude. Consider the vector plotted in Fig. 1.11. The vector is resolved by projecting the vector components onto the x and y axes. Thus, the projection of the vector along the x axis, V_x, is determined by fundamental trigonometric relations,

$$V_x = V \cos \theta,$$

and similarly the projection of the vector along the y axis is

$$V_y = V \sin \theta,$$

If given these components, one can work backwards and determine the

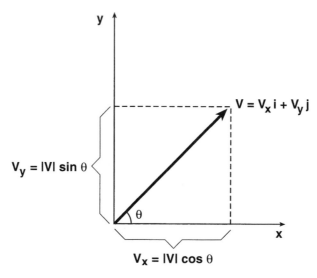

FIGURE 1.11. A vector **V** can be resolved in terms of its components with fundamental trigonometric relations.

magnitude and direction of the vector using

$$|\mathbf{V}| = \sqrt{V_x^2 + V_y^2},$$

$$\tan \theta = \frac{V_y}{V_x},$$

or

$$\theta = \tan^{-1}\left(\frac{V_y}{V_x}\right).$$

Thus, a vector can be reconstructed from its components.

1.6.2   Addition and Subtraction of Vectors

In many cases, a physical system under consideration undergoes several changes in either magnitude or direction within any given time interval. This involves the incorporation of several vectors, one for each corresponding change in the system. The objective in such a problem is to determine the total effect contributed by each individual change or vector. Given three vectors, **A**, **B**, and **C**, defined as

$$\mathbf{A} = 5\mathbf{i} + 3\mathbf{j} + 4\mathbf{k}, \quad \mathbf{B} = 15\mathbf{i} + \mathbf{j} + 6\mathbf{k}, \quad \mathbf{C} = 7\mathbf{i} + 10\mathbf{j} + 8\mathbf{k},$$

then the sum of the three vectors is the sum of the individual components from each of the three vectors:

$$\mathbf{A} + \mathbf{B} + \mathbf{C} = (5 + 15 + 7)\mathbf{i} + (3 + 1 + 10)\mathbf{j} + (4 + 6 + 8)\mathbf{k}$$

$$= 27\mathbf{i} + 14\mathbf{j} + 18\mathbf{k}.$$

Subtraction of a vector can be done by reversing the direction of the vector and adding to the remaining vectors. Reversing the direction of a vector simply changes the sign of the vector without affecting the magnitude. For example, subtracting vectors $\mathbf{B}$ and $\mathbf{C}$ from $\mathbf{A}$ is performed according to the following:

$$\mathbf{A} - \mathbf{B} - \mathbf{C} = (5\mathbf{i} + 3\mathbf{j} + 4\mathbf{k}) - (15\mathbf{i} + \mathbf{j} + 6\mathbf{k}) - (7\mathbf{i} + 10\mathbf{j} + 8\mathbf{k})$$

$$= (5 - 15 - 7)\mathbf{i} + (3 - 1 - 10)\mathbf{j} + (4 - 6 - 8)\mathbf{k}$$

$$= -17\mathbf{i} - 8\mathbf{j} - 10\mathbf{k}.$$

1.6.3 *Multiplication of Vectors*

Multiplication of vectors also serves an important role in the description of physical parameters. Given two vectors $\mathbf{A}$ and $\mathbf{B}$, the two methods of vector multiplication are as follows.[10]
  Scalar product:

$$\mathbf{A} \cdot \mathbf{B} = AB \cos \theta, \quad 0 \le \theta \le \pi$$

  Vector product:

$$\mathbf{A} \times \mathbf{B} = AB(\sin \theta)\mathbf{u}, \quad 0 \le \theta \le \pi$$

where $\mathbf{u}$ is a unit vector indicating the direction of $\mathbf{A} \times \mathbf{B}$. As an interesting note, the scalar product of two vectors is a scalar quantity while the vector product of two vectors is a vector quantity. Physical quantities derived from scalar products include the work (W) done by a force ($\mathbf{F}$) applied to a particle over a displacement ($\mathbf{d}$) ($W = \mathbf{F} \cdot \mathbf{d}$) and the potential energy (U) of an electric dipole in an external electric field ($\mathbf{E}$) given the electric dipole moment ($\mathbf{p}$) ($U = -\mathbf{p} \cdot \mathbf{E}$). Physical quantities derived from vector products are torque (τ), created by a force ($\mathbf{F}$) exerted at a vector position ($\mathbf{r}$) ($\tau = \mathbf{F} \times \mathbf{r}$), and the angular momentum ($\mathbf{L}$), generated by linear momentum ($\mathbf{p}$) exerted at a vector position ($\mathbf{r}$) ($\mathbf{L} = \mathbf{r} \times \mathbf{p}$).
  Two applications of vector analysis to problems of a physical nature include free-body diagrams of physical systems in static equilibrium and the representation of dynamic phenomena observed commonly in fluid flow problems. As an example of a free-body diagram related to the biological sciences, consider the example of an arterial bifurcation. Arterial bifurcations or branches, shown in Fig. 1.12, have been implicated in the development of cerebrovascular diseases due to their irregular geometry and hemo-

Arterial Bifurcation

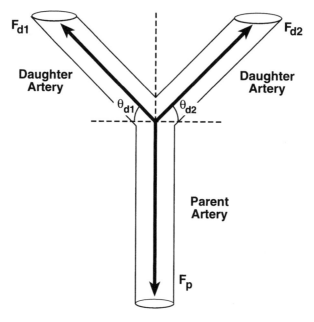

FIGURE 1.12. Illustrative application of vector analysis to an arterial bifurcation with the use of free-body diagrams.

dynamic instabilities. In fact, this issue it taken up in detail in the chapter on intracranial aneurysms (Chap. 6).

The point of importance in arterial bifurcations is the longitudinal stresses of the parent (main stem) and daughter (branches) arteries exerted on the apex. Each of the arteries involved in the bifurcation can be represented by vectors with a magnitude of force and a certain direction or angle with respect to the apex in the form of a free-body diagram (Fig. 1.12). The vectors are then resolved to reveal the total or resultant force acting in both the x and y directions. In static equilibrium, the only forces acting on the bifurcation apex are the forces of the arteries, F_p, F_{d1}, F_{d2}, which can be written in equation form as

$$F_p - F_{d1} - F_{d2} = 0.$$

These vectors can be resolved into their geometrical components along each axis. For the x axis,

$$(F_{d1})_x + (F_{d2})_x = 0,$$

$$F_{d1} \cos \Theta_1 + F_{d2} \cos \Theta_2 = 0.$$

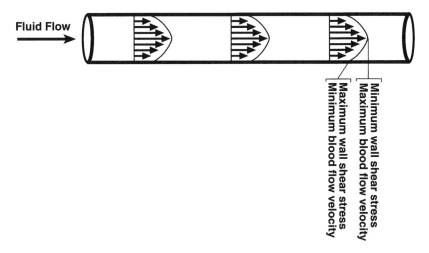

FIGURE 1.13. Vectors can be used to demonstrate hemodynamic effects of blood flow through a vessel. The direction of the vectors is always symmetric and parallel to the vessel longitudinal axis while the magnitude of the vectors correspond to the physical parameters blood flow velocity and wall shear stress, depending on their position within the velocity profile.

For the y axis,

$$(F_{d1})_y + (F_{d2})_y - (F_p)_y = 0,$$

$$F_{d1} \sin \Theta_1 + F_{d2} \sin \Theta_2 - F_p = 0.$$

If the magnitude of one of the forces is known, the two other forces can be determined by solving simultaneously the equation for the x and y axis.

The second application of vectors in this context is the characterization of biophysical phenomena observed commonly in fluid dynamics. Consider the laminar flow of a fluid within a rigid tube. Although more rigorous derivations of hemodynamically relevant parameters will be presented in Chap. 4, fluid flow exhibits a parabolic profile along the cross section of the tube. The profile is physically significant in that it resembles the distribution of flow velocity and wall shear stress, each of which can be represented by vectors (see Fig. 1.13). At the center of the profile where the vector is the longest, the flow velocity is largest. The flow velocity is minimal at the ends of the profile where the vectors are the smallest. The exact opposite is true for wall shear stress. The maximum wall shear stress occurs at the end of the profile or at the wall of the tube while it is at a minimum in the center of the profile.

1.7   Partial Differential Equations

Partial differential equations are differential equations containing partial derivatives. A partial derivative is the derivative of a three-dimensional function with respect to a certain parameter of interest while holding the other parameter constant. For example, given the three-dimensional function

$$z = x^2 y + 2yx + y^2 x,$$

then the partial derivatives of z with respect to x and y are, respectively,

$$\frac{\partial z}{\partial x} = 2xy + 2y + y^2 \quad \text{and} \quad \frac{\partial z}{\partial y} = x^2 + 2x + 2yx.$$

Probably the most important contribution from the partial derivative is not a particular equation but operators or mathematical directives that have properties similar in nature to algebraic equations. These operators act on functions or, more commonly, vectors to yield physically significant information concerning the behavior of the vector. More specifically, these two operators are the del operator,

$$\nabla \equiv \mathbf{i}\,\frac{\partial}{\partial x} + \mathbf{j}\,\frac{\partial}{\partial y} + \mathbf{k}\,\frac{\partial}{\partial z},$$

and the Laplacian operator,

$$\nabla^2 \equiv \frac{\partial^2}{\partial x^2} + \frac{\partial^2}{\partial y^2} + \frac{\partial^2}{\partial z^2}.$$

Each of these operators acts on the scalar and/or vector function by making the appropriate derivatives according to their individual components.

1.7.1 The Del Operator

The del operator can act on either a scalar quantity, where it now becomes known as a gradient, or a vector quantity. Assuming that Φ is a scalar function, then the gradient of Φ is determined by:

$$\text{grad } \Phi = \nabla\Phi = \left(\mathbf{i}\,\frac{\partial}{\partial x} + \mathbf{j}\,\frac{\partial}{\partial y} + \mathbf{k}\,\frac{\partial}{\partial z}\right)\Phi = \mathbf{i}\,\frac{\partial\Phi}{\partial x} + \mathbf{j}\,\frac{\partial\Phi}{\partial y} + \mathbf{k}\,\frac{\partial\Phi}{\partial z}.$$

With reference to a vector quantity, there are, in effect, two different operations that are possible. The first is the divergence of the vector $\mathbf{V}$ defined by

$$\text{div } \mathbf{V} = \nabla \cdot \mathbf{V} = \left(\mathbf{i}\,\frac{\partial}{\partial x} + \mathbf{j}\,\frac{\partial}{\partial y} + \mathbf{k}\,\frac{\partial}{\partial z}\right) \cdot (V_x \mathbf{i} + V_y \mathbf{j} + V_z \mathbf{k})$$

$$= \frac{\partial V_x}{\partial x} + \frac{\partial V_y}{\partial y} + \frac{\partial V_z}{\partial z}.$$

Each component of the gradient is multiplied by each component of the vector under the conditions that $\mathbf{i} \cdot \mathbf{i} = 1$ and cross products of the vector components such as $\mathbf{i} \cdot \mathbf{j} = 0$. The second operation is the curl of the vector

V, defined by

$$\text{curl } \mathbf{V} = \nabla \times \mathbf{V} = \left(\mathbf{i}\, \frac{\partial}{\partial x} + \mathbf{j}\, \frac{\partial}{\partial y} + \mathbf{k}\, \frac{\partial}{\partial z} \right) \times (V_x \mathbf{i} + V_y \mathbf{j} + V_z \mathbf{k})$$

$$= \left(\frac{\partial V_z}{\partial y} - \frac{\partial V_y}{\partial z} \right) \mathbf{i} + \left(\frac{\partial V_x}{\partial z} - \frac{\partial V_z}{\partial x} \right) \mathbf{j} + \left(\frac{\partial V_y}{\partial x} - \frac{\partial V_x}{\partial y} \right) \mathbf{k}.$$

Steps in the expansion of the curl of a vector were omitted purposefully at this stage but will be addressed later in the section on matrices (Sec. 1.8).

1.7.2 The Laplacian Operator

The Laplacian operator is similar to the gradient presented earlier with the exception that the Laplacian takes the second derivative of each component but can act on either a scalar function Φ or a vector function **V**:

$$\nabla^2 \Phi = \frac{\partial^2 \Phi}{\partial x^2} + \frac{\partial^2 \Phi}{\partial y^2} + \frac{\partial^2 \Phi}{\partial z^2},$$

$$\nabla^2 \mathbf{V} = \frac{\partial^2 V_x}{\partial x^2} \mathbf{i} + \frac{\partial^2 V_y}{\partial y^2} \mathbf{j} + \frac{\partial^2 V_z}{\partial z^2} \mathbf{k}.$$

The results correspond to the identity of an operating function, i.e., the Laplacian of a scalar is a scalar function and the Laplacian of a vector is a vector function. Each of these vector operations occurs in many areas of biophysics. The reader is urged to consult the references at the end of the chapter for further reading. As has been emphasized throughout this chapter, geometry plays a critical role in the expression of mathematical functions required to describe the biophysical phenomena involved in cerebrovascular diseases. These vector operators in rectangular, cylindrical, and spherical coordinate systems are summarized in Table 1.4. One may also extend these vector operations to more than two vectors. For example, given the cross product of two vectors **A** and **B**, $\mathbf{A} \times \mathbf{B}$, the cross product of another vector, **C**, can be performed with the original cross product by

$$(\mathbf{A} \times \mathbf{B}) \times \mathbf{C} = \mathbf{B}(\mathbf{A} \cdot \mathbf{C}) - \mathbf{A}(\mathbf{B} \cdot \mathbf{C}).$$

In such operations, the cross product encased within the parentheses is performed first, followed by the second cross-product operation. Such vector operations reveal a unique vector or scalar function that could also hold physical significance to the problem at hand. An ideal example will be seen in the discussion of the Navier–Stokes equation in Chap. 4.

Partial differential equations are used to characterize very important physical phenomena such as heat conduction, vibrating strings, and longitudinal and transverse vibrations of a beam. Solution of the majority of partial differential equations is a nontrivial matter, requiring extensive mathematical computation, and is beyond the scope of discussion. Inter-

TABLE 1.4. Vector and scalar operators expressed in rectangular, cylindrical, and spherical coordinates.

Rectangular coordinate system (x, y, z)

Del operator	$\mathbf{V} \equiv \mathbf{i} \dfrac{\partial}{\partial x} + \mathbf{j} \dfrac{\partial}{\partial y} + \mathbf{k} \dfrac{\partial}{\partial z}$
Gradient of a scalar	$\mathbf{V}\Phi = \mathbf{i} \dfrac{\partial \Phi}{\partial x} + \mathbf{j} \dfrac{\partial \Phi}{\partial y} + \mathbf{k} \dfrac{\partial \Phi}{\partial z}$
Divergence of a vector	$\mathbf{V} \cdot \mathbf{V} = \dfrac{\partial V_x}{\partial x} + \dfrac{\partial V_y}{\partial y} + \dfrac{\partial V_z}{\partial z}$
Curl of a vector	$\mathbf{V} \times \mathbf{V} = \left(\dfrac{\partial V_z}{\partial y} - \dfrac{\partial V_y}{\partial z} \right) \mathbf{i} + \left(\dfrac{\partial V_x}{\partial z} - \dfrac{\partial V_z}{\partial x} \right) \mathbf{j} + \left(\dfrac{\partial V_y}{\partial x} - \dfrac{\partial V_x}{\partial y} \right) \mathbf{k}$
Laplacian operator	$\mathbf{V}^2 \equiv \dfrac{\partial^2}{\partial x^2} + \dfrac{\partial^2}{\partial y^2} + \dfrac{\partial^2}{\partial z^2}$
Laplacian of a vector	$\mathbf{V}^2\mathbf{V} = \dfrac{\partial^2 V_x}{\partial x^2} \mathbf{i} + \dfrac{\partial^2 V_y}{\partial y^2} \mathbf{j} + \dfrac{\partial^2 V_z}{\partial z^2} \mathbf{k}$
Laplacian of a scalar	$\mathbf{V}^2\Phi = \dfrac{\partial^2 \Phi}{\partial x^2} + \dfrac{\partial^2 \Phi}{\partial y^2} + \dfrac{\partial^2 \Phi}{\partial z^2}$

Cylindrical coordinate system (r, ϕ, z)

Del operator	$\mathbf{V} \equiv \dfrac{\partial}{\partial r} \mathbf{r} + \dfrac{1}{r} \dfrac{\partial}{\partial \varphi} \boldsymbol{\phi} + \dfrac{\partial}{\partial z} \mathbf{z}$
Gradient of a scalar	$\mathbf{V}\Phi = \dfrac{\partial \Phi}{\partial r} \mathbf{r} + \dfrac{1}{r} \dfrac{\partial \Phi}{\partial \phi} \boldsymbol{\phi} + \dfrac{\partial \Phi}{\partial z} \mathbf{z}$
Divergence of a vector	$\mathbf{V} \cdot \mathbf{V} = \dfrac{1}{r} \left(\dfrac{\partial}{\partial r} (rV_r) + \dfrac{\partial V_\phi}{\partial \phi} + \dfrac{\partial V_z}{\partial z} \right)$
Curl of a vector	$\mathbf{V} \times \mathbf{V} = \left(\dfrac{1}{r} \dfrac{\partial V_z}{\partial \phi} - \dfrac{\partial V_\phi}{\partial z} \right) \mathbf{r} + \left(\dfrac{\partial V_r}{\partial z} - \dfrac{\partial V_z}{\partial r} \right) \boldsymbol{\phi}$ $+ \dfrac{1}{r} \left(\dfrac{\partial (rV_\phi)}{\partial r} - \dfrac{\partial V_r}{\partial \phi} \right) \mathbf{z}$
Laplacian operator	$\mathbf{V}^2 \equiv \dfrac{1}{r} \dfrac{\partial}{\partial r} \left(r \dfrac{\partial}{\partial r} \right) + \dfrac{1}{r^2} \dfrac{\partial^2}{\partial \phi^2} + \dfrac{\partial^2}{\partial z^2}$
Laplacian of a vector	$\mathbf{V}^2\mathbf{V} = \dfrac{1}{r} \dfrac{\partial}{\partial r} \left(r \dfrac{\partial V_r}{\partial r} \right) \mathbf{r} + \dfrac{1}{r^2} \dfrac{\partial^2 V_\phi}{\partial \phi^2} \boldsymbol{\phi} + \dfrac{\partial^2 V_z}{\partial z^2} \mathbf{z}$
Laplacian of a scalar	$\mathbf{V}^2\Phi = \dfrac{1}{r} \dfrac{\partial}{\partial r} \left(r \dfrac{\partial \Phi}{\partial r} \right) + \dfrac{1}{r^2} \dfrac{\partial^2 \Phi}{\partial \phi^2} + \dfrac{\partial^2 \Phi}{\partial z^2}$

Spherical coordinate system (r, θ, ϕ)

Del operator	$\mathbf{V} \equiv \dfrac{\partial}{\partial r} \mathbf{r} + \dfrac{1}{r} \dfrac{\partial}{\partial \theta} \boldsymbol{\theta} + \dfrac{1}{r \sin \theta} \dfrac{\partial}{\partial \phi} \boldsymbol{\phi}$
Gradient of a scalar	$\mathbf{V}\Phi = \dfrac{\partial \Phi}{\partial r} \mathbf{r} + \dfrac{1}{r} \dfrac{\partial \Phi}{\partial \theta} \boldsymbol{\theta} + \dfrac{1}{r \sin \theta} \dfrac{\partial \Phi}{\partial \phi} \boldsymbol{\phi}$
Divergence of a vector	$\mathbf{V} \cdot \mathbf{V} = \dfrac{1}{r^2} \dfrac{\partial}{\partial r} (r^2 V_r) + \dfrac{1}{r \sin \theta} \dfrac{\partial}{\partial \theta} ((\sin \theta) V_\theta) + \dfrac{1}{r \sin \theta} \dfrac{\partial V_\phi}{\partial \phi}$

TABLE 1.4 (*continued*)

Curl of a vector	$\mathbf{V} \times \mathbf{V} = \dfrac{1}{r^2 \sin\theta}\left(\dfrac{\partial\left(r(\sin\theta)V_\phi\right)}{\partial\theta} - \dfrac{\partial(rV_\theta)}{\partial\phi}\right)\mathbf{r}$
	$+ \dfrac{1}{r\sin\theta}\left(\dfrac{\partial V_r}{\partial\phi} - \dfrac{\partial\left(r\sin\theta)V_\phi\right)}{\partial r}\right)\theta + \dfrac{1}{r}\left(\dfrac{\partial(rV_\theta)}{\partial r} - \dfrac{\partial V_r}{\partial\theta}\right)\phi$
Laplacian operator	$\nabla^2 \equiv \dfrac{1}{r}\dfrac{\partial^2}{\partial r^2}(r^2) + \dfrac{1}{r^2\sin\theta}\dfrac{\partial}{\partial\theta}\left(\sin\theta\dfrac{\partial}{\partial\theta}\right) + \dfrac{1}{r^2\sin^2\theta}\dfrac{\partial^2}{\partial\phi^2}$
Laplacian of a vector	$\nabla^2\mathbf{V} = \dfrac{1}{r}\dfrac{\partial^2}{\partial r^2}\left(r^2\dfrac{\partial V_r}{\partial r}\right)\mathbf{r} + \dfrac{1}{r^2\sin\theta}\dfrac{\partial}{\partial\theta}\left(\sin\theta\dfrac{\partial V_\theta}{\partial\theta}\right)\theta + \dfrac{1}{r^2\sin^2\theta}\dfrac{\partial^2 V_\phi}{\partial\phi^2}\phi$
Laplacian of a scalar	$\nabla^2\Phi = \dfrac{1}{r}\dfrac{\partial^2}{\partial r^2}\left(r^2\dfrac{\partial\Phi}{\partial r}\right) + \dfrac{1}{r^2\sin\theta}\dfrac{\partial}{\partial\theta}\left(\sin\theta\dfrac{\partial\Phi}{\partial\theta}\right) + \dfrac{1}{r^2\sin^2\theta}\dfrac{\partial^2\Phi}{\partial\phi^2}$

ested readers are urged to consult the textbooks listed at the end of this chapter for additional reading.[11–13]

1.8   Matrices

Matrices are a rectangular array of numbers or elements characterized by its order. The order of a matrix, given as $p \times q$, reveals the overall size of the matrix by the number of elements in a row (p) and in a column (q). Therefore, given a 3×3 matrix,

$$\begin{bmatrix} a_{11} & a_{12} & a_{13} \\ a_{21} & a_{22} & a_{23} \\ a_{31} & a_{32} & a_{33} \end{bmatrix}$$

there are three rows, three columns, and nine elements of a_{ij}, where the subscript values indicate the position or location of the element within the matrix. The matrix above can be evaluated easily according to:

$$a_{11}(a_{22}a_{33} - a_{23}a_{32}) + a_{12}(a_{31}a_{23} - a_{33}a_{21}) + a_{13}(a_{21}a_{32} - a_{22}a_{31}).$$

The matrix is much more than an orderly display of numbers and plays a critical role in a variety of areas in the biophysical sciences including vector multiplication and image processing.

1.8.1   Vector Multiplication

In our discussion on vector multiplication, the expression for the curl of a vector **V** was stated without an explanation of the steps involved in the calculation and was deferred until the discussion of matrices. The reason for this deferment was that the particular steps involved in the calculation involved matrix analysis, making this section a more appropriate venue for discussion. The definition for the curl of a vector **V**,

$$\text{curl } \mathbf{V} = \nabla \times \mathbf{V} = \left(\mathbf{i} \frac{\partial}{\partial x} + \mathbf{j} \frac{\partial}{\partial y} + \mathbf{k} \frac{\partial}{\partial z} \right) \times \left(V_x \mathbf{i} + V_y \mathbf{j} + V_z \mathbf{k} \right),$$

can be expressed in matrix form according to the following:

$$\text{curl } \mathbf{V} = \begin{bmatrix} \mathbf{i} & \mathbf{j} & \mathbf{k} \\ \dfrac{\partial}{\partial x} & \dfrac{\partial}{\partial y} & \dfrac{\partial}{\partial z} \\ V_x & V_y & V_z \end{bmatrix}.$$

Expanding this matrix similar to the previous example,

$$\text{curl } \mathbf{V} = \mathbf{i} \begin{bmatrix} \dfrac{\partial}{\partial y} & \dfrac{\partial}{\partial z} \\ V_y & V_z \end{bmatrix} - \mathbf{j} \begin{bmatrix} \dfrac{\partial}{\partial x} & \dfrac{\partial}{\partial z} \\ V_x & V_z \end{bmatrix} + \mathbf{k} \begin{bmatrix} \dfrac{\partial}{\partial x} & \dfrac{\partial}{\partial y} \\ V_x & V_y \end{bmatrix}.$$

The two-dimensional matrices can be further evaluated, resulting in the following expression:

$$\text{curl } \mathbf{V} = \left(\frac{\partial V_z}{\partial y} - \frac{\partial V_y}{\partial z} \right) \mathbf{i} + \left(\frac{\partial V_x}{\partial z} - \frac{\partial V_z}{\partial x} \right) \mathbf{j} + \left(\frac{\partial V_y}{\partial x} - \frac{\partial V_x}{\partial y} \right) \mathbf{k}.$$

1.8.2   Image Processing

Optics, acoustics, and nuclear radiation are just a sample of the energy sources used in the generation of images. Regardless of the source, the generated images are all constructed in a similar manner. The images are, in essence, a matrix with the elements corresponding to a pixel. Each pixel represents a spatial distribution of the source energy with respect to the source position. For example, an image of a radioactive point source will demonstrate a localized region of increased intensity in the center of the image surrounded by a region of reduced intensity. The intensity within each pixel corresponds to the counts emitted by the source and registered by the radiation detector. With respect to cerebrovascular diseases and their small sizes, the diagnostic capabilities from an image become increasingly dependent on the image resolution. Increases in image resolution can be done to an extent by increasing the matrix order. Imaging techniques will be covered in more detail in each of the chapters pertaining to the diagnosis of the particular cerebrovascular diseases (Chap. 5, stroke; Chap. 6, aneurysms; and, Chap. 7, arteriovenous malformations). For additional reading on matrices, the reader should consult the textbooks referenced at the end of the chapter.[14–16]

1.9   Fourier Analysis

A number of biophysical and physiological processes possess the unique
characteristic of being cyclic or periodic in nature. Numerous examples can
be found in the human circulation that are, in turn, related to other periodic
functions. Consider, for example, the fluctuating variability of the pressures
experienced by the blood vessels during a cardiac cycle or a single heartbeat.
The blood pressure rises to a maximum at the onset of a heartbeat (systole)
corresponding to the initial surge of blood from the heart into the blood
vessels feeding the entire body. The blood pressure reaches a minimum at
cardiac diastole before returning to baseline values of blood pressure. The
periodic fluctuations of the blood pressure translate into periodic behavior
of other parameters that are dependent on the pressure, including flow, and
the stress and strain of the blood vessels. Although this cycle is repeated
approximately 60–70 times per minute, the pressure values and, in some
cases, the shape of the pressure curve are subject to variational disturbances
brought about by various levels of daily activity and preexisting health conditions such as arterial hypertension.

Due to their periodic nature, these hemodynamic (pressure and flow)
waveforms can be characterized mathematically by the Fourier series representation of a periodic function $f(x)$. By expanding a function into a
Fourier series, one can represent mathematically any complex waveform by
calculating harmonics. Assume $f(t)$ is defined over an interval defined by
the range $-T/2$ to $T/2$ and has a period of T. Then, $f(t)$ can be expressed
according to the following[17]:

$$f(t) = A_0 + \sum_{n=1}^{\infty} \left[A_n \cos\left(\frac{2\pi n}{T}\right)t + B_n \sin\left(\frac{2\pi n}{T}\right)t \right].$$

The Fourier coefficients A_0, A_n, and B_n in $f(t)$ are

$$A_0 = \frac{1}{T} \int_{-T/2}^{T/2} f(t)dt,$$

$$A_n = \frac{2}{T} \int_{-T/2}^{T/2} f(t) \cos\left(\frac{2\pi n t}{T}\right) dt,$$

$$B_n = \frac{2}{T} \int_{-T/2}^{T/2} f(t) \sin\left(\frac{2\pi n t}{T}\right) dt,$$

where $n = 0, 1, 2, 3, \ldots$ The physical significance of the integers when calculating the Fourier series is that they represent the harmonics of the periodic
waveform. The case of $n = 1$ corresponds to the first or fundamental harmonic, while subsequent integer values reveal higher harmonics.

Hemodynamic waveforms are unique in amplitude, shape, and frequency
according to vessel type and the point or position along the particular vessel

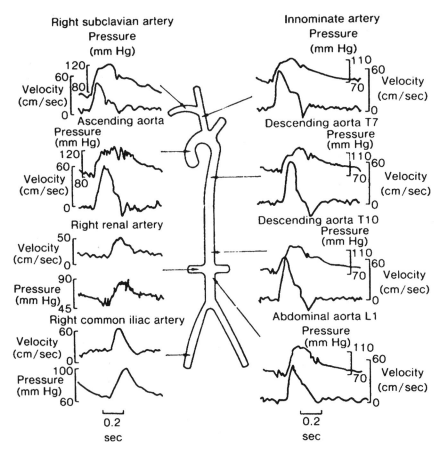

FIGURE 1.14. Hemodynamic waveforms of pressure and flow vary according to their position or location within the human aorta. (Reprinted from *Cardiovascular Research* **4**, C.J. Mills, I.T. Gabe, J.H. Gault, D.T. Mason, J. Ross, Jr, E. Braunwald, and J.P. Shillingford, Pressure–flow relationships and vascular impedance in man, pp. 405–417, 1970, with kind permission of Elsevier Science–NL, Sara Burgerhartstraat 25, 1055 KV Amsterdam, The Netherlands.)

(Fig. 1.14). Although they are different in comparison to any two people, quantitative analysis of hemodynamic waveforms can accurately demonstrate hemodynamic abnormalities consequent to cerebrovascular diseases such as vessel occlusion or stenosis observed in occlusive artery disease (Chap. 5) and the rapid shunting of blood exhibited by arteriovenous malformations (Chap. 7). Quantitative analysis of any type of hemodynamic (pressure or flow) waveforms can be accomplished through the mathematical characterization of the waveforms using Fourier analysis (Fig. 1.15). The applications of Fourier analysis to the biophysical sciences are many and are described in detail in several textbooks.[18-22]

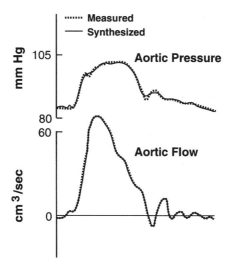

FIGURE 1.15. Application of Fourier analysis in the characterization of hemodynamic waveforms. (From: Nichols WW, O'Rourke MF. *McDonald's Blood Flow in Arteries*. Philadelphia: Lea & Febiger; 1990: p. 276. © Williams & Willkins Reprinted with permission.)

1.10   Complex Variables

Complex variables include any mathematical expression that involves the complex number $i = (-1)^{1/2}$ and is presented typically in the form

$$z = x + iy,$$

where z is a complex variable, x is a real coefficient, y is a complex coefficient, and i is a complex number. When presented with the concept of a complex variable or number, one finds it difficult intuitively to comprehend the meaning of a complex variable or number and subsequently becomes hard-pressed to find any possible applications to physical phenomena, much less to make the extension to applications of cerebrovascular diseases. In fact, complex variables constitutes the entire range of integers (considering the case where $y = 0$) and can be found in many areas of physics, including kinematics and the basic equations of motion.[23]

Consider the example of a child, Alexandra, who is on a nature hunt and spots a butterfly at rest on a plant 20 meters away. Alexandra begins her pursuit of the butterfly, running at a maximum speed 4 m/s. However, at that instant, the butterfly senses danger and flutters off at an acceleration of 2 m/s^2. The question to be answered is whether she will capture the butterfly or whether the butterfly will escape. Also in question is the speed that Alexandra would need to maintain to have any hopes of catching the butterfly. This solution can be obtained easily by analyzing the motion of

Alexandra and the butterfly and determining if and at what time their position is equal. Using the standard equations of motion, the position of Alexandra, x_A, is given by

$$x_A = v_0 t = (4 \text{ m/s})t$$

and the position of the butterfly, x_b, given by

$$x_b = x_0 + \frac{at^2}{2} = 20 \text{ m} + (1 \text{ m/s}^2)t^2.$$

Setting these two equations of motion equal to each other and solving for time yields

$$(4 \text{ m/s})t = 20 \text{ m} + (1 \text{ m/s}^2)t^2$$

or

$$20 - 4t + t^2 = 0.$$

The solution of a quadratic equation, i.e., $ax^2 + bx + c = 0$, is given by

$$x = \frac{-b \pm \sqrt{b^2 - 4ac}}{2a},$$

where, for this problem, $a = 1$, $b = -4$, and $c = 20$. Substituting these values into the equation above yields the time at which the position of Alexandra coincides with that of the butterfly:

$$x = \frac{-(-4) \pm \sqrt{(-4)^2 - 4(1)(20)}}{2(1)}$$

$$= \frac{4}{2} \pm \frac{\sqrt{16 - 80}}{2} = 2 \pm \frac{\sqrt{-64}}{2} = 2 \pm i\sqrt{16},$$

where x represents time. The complex component in the answer leads one to wonder if and to what extent this answer can be applied to the problem at hand. According to a similar problem demonstrated by Newburgh,[23] the time of closest approach between Alexandra and the butterfly is represented by the real part of the complex expression above, i.e., $t = 2$ s, and that the velocity of Alexandra required to capture the butterfly is calculated from the equation of motion $v_0 = 2ax$ and found to be 8.9 m/s. Thus, although it may be confusing initially, physically realistic information can be extracted readily from complex expressions.

From the above definition, a complex variable can be further expanded into trigonometric components according to the folowing relation:

$$z = x + iy = R(\cos\theta + i\sin\theta),$$

where R and θ are, in essence, polar coordinates but plotted on a graph with the x axis representing the real component and the y axis representing the

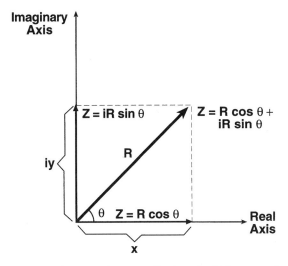

FIGURE 1.16. Schematic diagram of a vector **R** plotted within a complex plane where the ordinate is the imaginary axis and the abscissa is the real axis.

complex component (Fig. 1.16). From a trigonometric standpoint, it can be seen readily that substituting $\theta = 0°$ gives $z = R$ and $\theta = 90°$ gives $z = iR$, implying a rotation of the vector R by $90°$. Since the complex exponential function can be expressed in terms of the periodic trigonometric functions, then it becomes possible to define waves as complex entities:

$$\psi = Ae^{i(\omega t + \phi)},$$

which can also be expressed in terms of cosine and sine terms:

$$\phi = A\cos(\omega t + \phi) + iA\sin(\omega t + \phi)$$

The angular frequency ω and phase angle ϕ are constant throughout the propagation of the wave. Thus, as the time t increases, the wave travels in a periodic nature as a rotating vector in the imaginary plane. The complex description of a traveling wave will be important in describing the periodic pressure and flow waveforms as a result of pulsatile blood flow. Additional sources of information on complex variables are cited for interested readers.[24–27]

1.11 Summary

Biological sciences, particularly in applications of physiology, are an excellent example of where a multitude of mathematical disciplines and topics converge to provide an in-depth qualitative and quantitative study of the complex processes involved in all aspects of cerebrovascular disease. The purpose of this chapter was to introduce these mathematical topics, primarily through a brief treatment of theory combined with illustrative examples. In doing so, it becomes easy to appreciate their potential role in the description of physiological processes in cerebrovascular disease. However, in an attempt to circumvent any possible confusion between the mathematics and the physics involved in cerebrovascular diseases, a concerted effort was made, following the discussion on a given mathematical concept, to draw attention and direct the reader to relevant biophysical phenomena in subsequent chapters.

1.12   References

1. D. Halliday and R. Resnick, *Fundamentals of Physics*, 2nd ed. (Wiley, New York, 1981), pp. 44–46.
2. Z.X. Wang and D.R. Guo, *Special Functions* (World Scientific, Singapore, 1989).
3. B.C. Carlson, *Special Functions of Applied Mathematics* (Academic, New York, 1977).
4. M. Abramowitz and I. Stegun, *Handbook of Mathematical Functions* (Dover, New York, 1965).
5. E. Jahnke, F. Emde, and F. Losch, *Tables of Higher Functions* (McGraw-Hill, New York, 1960).
6. C. Isenberg and S. Chomet, *Physics Experiments and Projects for Students*, (Newman-Hemisphere, London, 1985), Vol. I, pp. 46–47.
7. J.C. Burkhill, *The Theory of Ordinary Differential Equations* (Longman, New York, 1975).
8. G. Birkhoff and G. Rota, *Ordinary Differential Equations*, 4th ed. (Wiley; New York, 1989).
9. J.R. Acton and P.T. Squire, *Solving Equations with Physical Understanding* (Hilger, Bristol, 1985).
10. M.R. Spiegel, *Schaum's Outline of Theory and Problems of Advanced Mathematics for Engineers and Scientists* (McGraw-Hill, New York, 1971).
11. A. Sommerfeld, *Partial Differential Equations in Physics* (Academic, New York, 1967).
12. F. John, *Partial Differential Equations* (Springer, New York, 1982).
13. R. Dennermeyer, *Introduction to Partial Differential Equations and Boundary Value Problems* (McGraw-Hill, New York, 1968).
14. H. Eves, *Elementary Matrix Theory* (Allyn and Bacon, Boston, 1966).
15. S. Barnett, *Matrices: Methods and Applications* (Oxford University Press, New York, 1990).
16. S. Barnett, *Matrix Methods for Engineers and Scientists* (McGraw-Hill, New York, 1979).
17. E.O. Attinger, A. Anne, and D.A. McDonald, "Use of Fourier series for the analysis of biological systems", Biophys. J. **6**, 291–304 (1966).
18. R.V. Churchill, and J.W. Brown, *Fourier Series and Boundary Value Problems*, 3rd ed. (McGraw-Hill, New York, 1978).

19. M.R. Spiegel, *Schaum's Outline of Theory and Problems of Fourier Analysis with Applications to Boundary Value Problems* (McGraw-Hill, New York, 1974).
20. R.N. Bracewell, *The Fourier Transform and its Applications*, 2nd ed., revised (McGraw-Hill, New York, 1986).
21. R.N. Bracewell, "The Fourier transform," Sci. Am. **260**, 86–89, 92–95 (1989).
22. R.E. Challis and R.I. Kitney, "Biomedical signal processing. Part 2. The frequency transforms and their interrelationships," Med. Biol. Eng. Comput. **29**, 1–17 (1991).
23. R. Newburgh, "Real, imaginary, and complex numbers: Where does the physics hide?" Phys. Teacher **34**, 23–25 (1996).
24. G. Moretti, *Functions of a Complex Variable* (Prentice-Hall, Englewood Cliffs, NJ, 1964).
25. G.F. Carrier, M. Krook, and C.E. Pearson, *Functions of a Complex Variable* (McGraw-Hill, New York, 1966).
26. R.V. Churchill, J.W. Brown, and R.F. Verhey, *Complex Variables and Applications* (McGraw-Hill, New York, 1974).
27. E.T. Copson, *Theory and Functions of a Complex Variable* (Oxford University Press, New York, 1962).

1.13   Problems

1.1.   Classify the following functions in terms of their classification: (A) $y = 0$, (B) $y = x$, (C) $y = e^{2x} + 6$, (D) $y = \sin(x^2)$, and (E) $y = \ln[\sin(x)]$.

1.2.   What would be the slope of the function $z = \sin(x)\cos(y)$?

1.3.   Express the three directional coordinates of both (A) cylindrical (r, ϕ, z) and (B) spherical (r, θ, ϕ) coordinates in terms of rectangular coordinates.

1.4.   What is the derivative of the y intercept of a function?

1.5.   Given a particle whose position in space is defined by the vector $x = A\sin(\omega t)\mathbf{i} + A\cos(\omega t)\mathbf{j}$. (A) What is the particle's distance from the origin? (B) What is the velocity of this particle? (C) What is the speed and direction of the particle? (D) What is the acceleration of the particle? (E) What is the general direction of the acceleration of the particle with respect to its velocity? Explain.

1.6.   Radioactive elements are typically characterized by a decay constant in terms of their half-life $(k_{1/2})$. Derive the decay constant assuming a quarter-life or $k_{1/4}$.

1.7.   Repeat the derivation of the equation for radioactive decay incorporating the rate of production, R, of the radioactive species upon decay of the original radioactive element.

1.8.   If the photon energy of a 1-millicurie (mCi) sample of ^{99m}Tc is 140 keV at $t = 0$, what would the energy of a photon emitted from the sample at its half-life, $t = 6$ hours, be?

1.9.   What is the projection of $\psi = Ae^{i(\omega t + \phi)}$ (A) along the real axis and (B) along the complex or imaginary axis?

1.10. We have an inductance–capacitance–resistance (LCR) electrical circuit described by the differential equation

$$L\frac{d^2q}{dt^2} + R\frac{dq}{dt} + \frac{q}{C} = V_0\cos(\omega t),$$

where L is the inductance, q the electrical charge, R the resistance, C the capacitance, and V_0 the voltage. Using the general form of

$$q = q_0\cos(\omega t + \phi),$$

where q_0 is the maximum charge of the capacitor, find the general solution of the differential equation.

1.11. Express the given pressure wave function $P = Ae^{i\phi}e^{2\pi\omega t}$ in terms of trigonometric functions.

1.12. Given a periodic wave described by $P = ke^{i\omega t}$, what is the derivative of this function with respect to time? With respect to the frequency?

1.13. Assuming that nodal and loop analysis of a particular electrical network revealed the following equations:

$$\text{Nodal}: \quad i_i = i_2 + i_3.$$

$$\text{Loop}: \quad V_1 - i_1R_1 + V_2 - i_2R_2 = 0,$$

$$V_2 - i_2R_2 - V_3 + i_3R_3 = 0.$$

Use matrix analysis to set up the solution for the current.

1.14. The Gaussian curve is used in many applications to the physical sciences ranging from radioactive decay to image processing to statistical physics and is given by $y = e^{-A^2x^2}$, where A is a constant with units inverse to that of x. The shape of the Gaussian curve resembles a symmetric bell curve. An important feature of the Gaussian curve or distribution is a quantitative measure of the full width at half maximum (FWHM), which reflects the spread of the curve in the central portion of the function. Derive an expression for the FWHM.

2
The Physics of Elasticity

2.1 Introduction

Regardless of its structural composition, an object acted on by external forces such as pushing, pulling, twisting, bending, or stretching will respond to the forces through a deformation or change in the size and/or shape of the object. This is guaranteed by Newton's third law, which states that for every force acting on an object, the object responds with an equal yet opposite reacting force. The magnitude of force required to elicit such a response depends on the object's internal structure and size. However, what is of interest is not how the object responds during the exertion of these forces, but how and why the object responds once these forces are removed. In other words, is the deformation that occurs as the result of the external forces temporary or permanent and, if permanent, to what extent is the object deformed? These questions form the physical basis for elasticity.

Elasticity is not a measure of the ability to stretch or distort because of externally applied forces, but the ability to return to the original form (size and shape) when the distorting force is removed.[1] Elasticity is characterized by the presence and degree of deformation and depends primarily on the size and internal structure of the object, the magnitude of the force, and the time interval or rate over which the force is applied. The presence of deformation can be determined through visual observation, but how is the degree of deformation defined? This can be clarified through a simple demonstration with commonly found materials: a rubber band and modeling clay.

A rubber band can be subjected to external forces such as stretching, twisting, pulling, pushing, and bending for lengthy time intervals and still revert to its original size and shape once these forces are removed. Modeling clay, on the other hand, assumes and maintains a deformed state as a result of these applied forces. The elasticity of a substance can be characterized mathematically by the plot of stress versus strain. The stress and strain of elastic objects will be covered in detail in the following sections. For the purpose of this discussion, the stress reflects the magnitude of force required

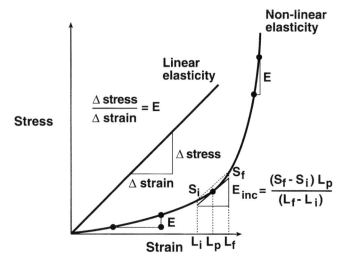

FIGURE 2.1. Representative stress–strain curves for a linear and nonlinear elastic object. The slope of the stress–strain curve is the elastic modulus. The elastic modulus at the curved portion of the nonlinear stress–strain curve is approximated by $L_p = (L_i + L_f)/2$. The approximation for L_p becomes more accurate as the difference between L_i and L_f becomes smaller.

to induce the given deformation and strain represents the fractional response of the deformation due to the stress. The rubber band exhibits a primarily linear relationship between stress and strain and is thus considered a linearly elastic substance, at least to a first-order approximation. The stress and strain curve of modeling clay, which is an inelastic object, is linear only for the brief beginning of the curve and increases rapidly. Figure 2.1 shows representative stress–strain curves for a linearly and nonlinearly elastic object. The rapid increase in strain as a result of stress implies that modeling clay is a predominantly nonlinear substance.

Elasticity is an important physical factor in the anatomy and physiology of blood vessels of the human body and in many developmental aspects of cerebrovascular diseases. The elasticity of the vessel wall is influenced by physiological processes such as aging, hypertension, and atherosclerosis and is adversely affected at each stage of the disease process. In short, vessel wall elasticity plays a critical role in the course of development and ultimate fate of cerebrovascular diseases. This chapter will address the major topics and terms of elasticity as they pertain to the biophysical treatment and discussion of cerebrovascular diseases covered in later chapters. More specifically, the following concepts comprising the physics of elasticity will be described: (1) a physical and mathematical definition of the mechanical stresses and strains that act on solids and the physical characterization of elastic properties of solids using elastic moduli such as Young's modulus, bulk modulus, and shear modulus in addition to the elastic limit; (2) the introduction and application of Hooke's law and Laplace's law to describe the statics and dynamics of an elastic solid; (3) the introduction and application of viscoelasticity in describing the elasticity of biological materials; and (4) an elementary treatment of the physics of waves and wave propagation.

2.2   Stress

The application of an external force to an elastic material results in some degree of deformation. The magnitude or strength of the external force causing the deformation is defined as the physical quantity, stress. In physical terms, stress is equal to the force F divided by the surface area of the elastic material or

$$S = \frac{F}{A} \tag{2.1}$$

and is given in units of dyn/cm^2. It is the stresses that are of primary concern to seismologists when analyzing a fault region of the earth for the propensity of earthquakes. In a fault region, developed plates within the earth's core are positioned against each other and remain dormant in equilibrium. However, shifting of the plates, which represents the occurrence of minor earthquakes, results in the development and accumulation of stresses in other fault regions in close proximity to the current fault. The distribution and magnitude of stresses that occur as a direct result of an earthquake are currently critical issues of research in the seismological prediction and detection of major earthquakes.

Another illustrative example of stress is the physical behavior of a spring. Applying an external force by stretching the spring temporarily deforms the spring and, at equilibrium, creates a resistive force or stress that is equal in magnitude and acts in a direction opposite to the applied force. Once the applied force is removed, the stresses are released, causing the spring to oscillate before ultimately returning to its original shape prior to the deformation. More importantly, energy created from the external force in stretching the spring is transferred and stored in the spring during the deformation of the spring. Releasing the spring results in a transfer of this energy to motion before dissipation from a frictional resistance due to air (gravity).

Consider the elastic cube presented in Fig. 2.2(A). If left alone without the presence of external forces, the elastic cube does not move and maintains its size and shape. The cube is in static equilibrium. When subjected to an external force such as poking a finger through any face of the cube, pulling opposite sides of the cube along a longitudinal direction, or thrusting a palm

TYPE OF ELASTICITY	VISUAL DESCRIPTION	ELASTICITY RELATIONS

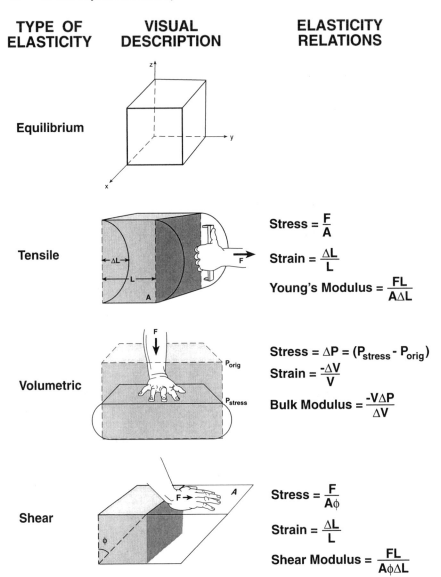

Equilibrium

Tensile

$$\text{Stress} = \frac{F}{A}$$

$$\text{Strain} = \frac{\Delta L}{L}$$

$$\text{Young's Modulus} = \frac{FL}{A\Delta L}$$

Volumetric

$$\text{Stress} = \Delta P = (P_{stress} - P_{orig})$$

$$\text{Strain} = \frac{-\Delta V}{V}$$

$$\text{Bulk Modulus} = \frac{-V\Delta P}{\Delta V}$$

Shear

$$\text{Stress} = \frac{F}{A\phi}$$

$$\text{Strain} = \frac{\Delta L}{L}$$

$$\text{Shear Modulus} = \frac{FL}{A\phi\Delta L}$$

FIGURE 2.2. Schematic diagram of an elastic object: (A) in equilibrium, (B) subjected to a longitudinal or tensile stress, (C) subjected to a volumetric stress, and (D) subjected to a shearing stress.

along the top of the cube, all planes, all sides, all surface elements, and each three-dimensional point within the cube are subjected to an elastic force(s) in response to the external force(s) exerted by the hand. This elastic force, known as stress, is the physical equivalent to pressure acting on a solid object and represents force per unit area. In brief, there exist primarily three types of stresses on an object: longitudinal stress, volumetric stress, and shear stress. Longitudinal or tensile stress occurs as a result of equal yet opposite forces applied to two opposing sides of the cube and is demonstrated in Fig. 2.2(B). Tensile stress is a stress created by an external force exerted in a direction perpendicular or normal to the surface area of a particular plane of the cube (S_{xx}, S_{yy}, S_{zz}). Volumetric stress results from the application of a uniform compressive force along the surface area of the cube and is demonstrated in Fig. 2.2(C) by pounding any face of the cube. Shear stress is the result of a rotational force or torque caused by equal yet opposite forces along a specific plane or axis and is demonstrated in Fig. 2.2(D) by the thrusting of the palm in a parallel direction along the top of the cube. Shearing stress is a stress that acts in a direction parallel to the surface area of the plane to which the external force was applied (S_{xy}, S_{yz}, S_{xz}). With regard to the nine possible stress components of an elastic object, six components are used to describe the elasticity of the cube with three components eliminated due to symmetry, i.e., $S_{xy} = S_{yx}$, $S_{yz} = S_{zy}$, and $S_{xz} = S_{zx}$.

Assuming that the cube is centered within the coordinate plane defined by x, y, and z, each point or plane within the cube can be represented by proper assignment of the coordinate axes, x, y, z. The stress at a point within the cube combines all of the forces acting on surface elements that contain the point, whatever their orientation, but it is convenient analytically to resolve a given stress into three orthogonal components.[2] Thus, assuming that the cube remains stationary, i.e., no rotation of the cube, there exist six different components of stress: S_{xx}, S_{yy}, S_{zz}, S_{xy}, S_{xz}, S_{yz}. Since stress is a three-dimensional quantity, each directional force F acting on any of the cube surfaces parallel to each coordinate axis can be expressed mathematically in terms of three components of stress: one component of tensile or volumetric stress and two components of shearing stress:

$$|\mathbf{F}_x| = \mathbf{i}S_{xx} + \mathbf{j}S_{xy} + \mathbf{k}S_{xz},$$
$$|\mathbf{F}_y| = \mathbf{i}S_{yx} + \mathbf{j}S_{yy} + \mathbf{k}S_{yz}, \tag{2.2}$$
$$|\mathbf{F}_z| = \mathbf{i}S_{zx} + \mathbf{j}S_{zy} + \mathbf{k}S_{zz},$$

where $\mathbf{i}$, $\mathbf{j}$, $\mathbf{k}$ are unit vectors in the direction of the coordinate system. The resultant force $\mathbf{F}_{res}$, acting on a surface element of the elastic cube is given as

$$F_{res} = \frac{\partial F_x}{\partial x} + \frac{\partial F_y}{\partial y} + \frac{\partial F_z}{\partial z}.$$

The stress of the elastic material due to volumetric compression is given by

$$S = -\tfrac{1}{3}\left(S_{xx} + S_{yy} + s_{zz}\right). \tag{2.3}$$

2.3 Strain

Another physical quantity used to describe the elasticity of a solid is strain. Returning to our example of the elastic cube, assume that stresses are applied such that the cube experiences a permanent deformation. Strain, in effect, is a quantitative measure of the fractional extent of deformation of the elastic cube produced as a result of the applied stresses and is a unitless quantity. Strain is defined as the ratio of the increase in a particular dimension in the deformed state to that dimension in its initial, undeformed state[3] and can be expressed mathematically as

$$\varepsilon = \frac{x_{\text{def}} - x_{\text{nor}}}{x_{\text{nor}}}, \tag{2.4}$$

where ε is strain and x_{def} and x_{nor} represent the positional coordinates of the elastic object in the deformed and normal states, respectively. Each component of strain is related to each component of stress, and there is no simple pairing of the various components in the stress–strain relationship.[4] As the strain within an elastic object is dependent on the applied stress, the types of strain are similar in nature to the types of stress. For example, a tensile or longitudinal stress produces a tensile strain again acting in a direction perpendicular or normal to the surface area of a particular plane of the cube $(\varepsilon_{xx}, \varepsilon_{yy}, \varepsilon_{zz})$ to which it was applied and is defined mathematically by

$$\varepsilon_{xx} = \left(\frac{\Delta L}{L}\right)_x, \quad \varepsilon_{yy} = \left(\frac{\Delta L}{L}\right)_y, \quad \varepsilon_{zz} = \left(\frac{\Delta L}{L}\right)_z, \tag{2.5}$$

where $\varepsilon_{xx}, \varepsilon_{yy}, \varepsilon_{zz}$ are the normal strain components acting in the unit coordinate directions $\mathbf{i}, \mathbf{j}, \mathbf{k}$, and $(\Delta L/L)_x, (\Delta L/L)_y, (\Delta L/L)_z$ represent the changes in length produced by the deformation along the corresponding three coordinate directions. Assuming that a given displacement p is experienced by the elastic cube as a result of an applied stress, then the normal strains can be expressed mathematically as

$$\varepsilon_{xx} = \frac{\partial p_x}{\partial x}, \quad \varepsilon_{yy} = \frac{\partial p_y}{\partial y}, \quad \varepsilon_{zz} = \frac{\partial p_z}{\partial z}.$$

In a similar manner, the shearing or angular strain, which is a strain that results from a stress acting in a direction parallel to the surface area of the plane to which the external force was applied (ε_{xy}, ε_{yz}, ε_{xz}), is given by

$$\varepsilon_{xy} = \frac{1}{2}\left(\frac{\partial p_x}{\partial y} + \frac{\partial p_y}{\partial x}\right), \quad \varepsilon_{yz} = \frac{1}{2}\left(\frac{\partial p_y}{\partial z} + \frac{\partial p_z}{\partial y}\right), \quad \varepsilon_{xz} = \frac{1}{2}\left(\frac{\partial p_x}{\partial z} + \frac{\partial p_z}{\partial x}\right).$$

As has been stressed in previous sections, geometry is an important issue in the description of biophysical concepts, particularly as they pertain to cerebrovascular diseases. Thus, the strains can be expressed in cylindrical and spherical coordinates according to the following relations.

Cylindrical coordinates:

$$\varepsilon_{rr} = \frac{\partial u_r}{\partial r},$$

$$\varepsilon_{\phi\phi} = \frac{1}{r}\frac{\partial u_\phi}{\partial \phi} + \frac{u_r}{r},$$

$$\varepsilon_{zz} = \frac{\partial u_z}{\partial z},$$

$$\varepsilon_{\phi z} = \frac{1}{r}\frac{\partial u_z}{\partial \phi} + \frac{\partial u_\phi}{\partial z},$$

$$\varepsilon_{zr} = \frac{\partial u_r}{\partial z} + \frac{\partial u_z}{\partial r},$$

$$\varepsilon_{r\phi} = \frac{\partial u_\phi}{\partial r} - \frac{u_\phi}{r} + \frac{1}{r}\frac{\partial u_r}{\partial \phi}.$$

Spherical coordinates:

$$\varepsilon_{rr} = \frac{\partial u_r}{\partial r},$$

$$\varepsilon_{\theta\theta} = \frac{1}{r}\frac{\partial u_\theta}{\partial \theta} + \frac{u_r}{r},$$

$$\varepsilon_{\phi\phi} = \frac{1}{r\sin\theta}\frac{\partial u_\phi}{\partial \phi} + \frac{u_\theta}{r}\cot\theta + \frac{u_r}{r},$$

$$\varepsilon_{\theta\phi} = \frac{1}{r}\left(\frac{\partial u_\phi}{\partial \theta} - u_\phi\cot\theta\right) + \frac{1}{r\sin\theta}\frac{\partial u_\theta}{\partial \phi},$$

$$\varepsilon_{\phi r} = \frac{1}{r\sin\theta}\frac{\partial u_r}{\partial \phi} + \frac{\partial u_\phi}{\partial r} - \frac{u_\phi}{r},$$

$$\varepsilon_{r\theta} = \frac{\partial u_\theta}{\partial r} - \frac{u_\phi}{r} + \frac{1}{r}\frac{\partial u_r}{\partial \theta}.$$

The longitudinal strains (ε_{xx}, ε_{yy}, ε_{zz}) are related to the shearing or transverse strains by Poisson's ratio σ, which is unitless and is defined according to

$$-\sigma_{xy} = \frac{\varepsilon_{xx}}{\varepsilon_{yy}}, \quad -\sigma_{xz} = \frac{\varepsilon_{xx}}{\varepsilon_{zz}}, \quad -\sigma_{yz} = \frac{\varepsilon_{yy}}{\varepsilon_{zz}},$$

$$-\sigma_{yx} = \frac{\varepsilon_{yy}}{\varepsilon_{xx}}, \quad -\sigma_{zx} = \frac{\varepsilon_{zz}}{\varepsilon_{xx}}, \quad -\sigma_{zy} = \frac{\varepsilon_{zz}}{\varepsilon_{yy}}.$$

Poisson's ratio for human blood vessels is often approximated by 0.5.[5] A value of 0.5 implies that the volume of the substance remains constant as it is elongated, a relationship that can be derived from the above equations for Poisson's ratio and the definition of strain.[6]

The final type of strain is the volumetric strain, a uniform compression of an elastic solid along one projection causing a volumetric deformation. The volumetric strain, according to previous definitions, is given by the change of volume along a given projection or

$$\varepsilon_{xx} = \left(\frac{\Delta V}{V}\right)_x, \quad \varepsilon_{yy} = \left(\frac{\Delta V}{V}\right)_y, \quad \varepsilon_{zz} = \left(\frac{\Delta V}{V}\right)_z. \tag{2.6}$$

The physical interactions of elasticity and their corresponding equations described above are summarized in Table 2.1.

2.4   Hooke's Law

Any material that is linearly elastic, i.e., restored to its original form following the application of an external force, exhibits a linear relation between the tensile stress and tensile strain. In fact, the stress–strain relationships of most materials are linear toward the beginning of the curve before the curve becomes nonlinear. The linear relation between stress and strain is referred to as Hooke's law. Materials that conform to Hooke's law and behave in a manner such that stress is proportional to strain are known as Hookean solids. The proportionality constant, also represented by the slope or derivative of the linear portion of the stress–strain curve, yields a modulus of elasticity known as Young's modulus. The elastic modulus possesses the same physical units as that of stress (dyn/cm^2). Since the elastic modulus is linearly proportional to stress, then a large value for the elastic modulus implies that a large stress is required to produce a given strain. Consider, as a typical example, steel, which has a very high elastic modulus (2×10^{12} dyn/cm^2) as one would expect. Common experience indicates that an extremely large force as stress would need to be applied to a steel object in order to induce a small deformation. The detailed description and a discussion of the various types of elastic moduli will be covered in the following section.

This concept, however, cannot be extended to nonlinearly elastic materials. Referring to the graph of the nonlinear modeling clay, the stress–strain

TABLE 2.1. Physical interactions of elasticity.

Tensile elasticity

Tensile stress: $\dfrac{F}{A}$ (N/m^2)

F = exerted force causing the stress
A = cross-sectional area upon which the force acts

Tensile strain: $\dfrac{\Delta L}{L}$

ΔL = change in length as a result of stress (deformed − original)
L = original length of material

Young's modulus: $Y = \dfrac{(\text{tensile stress})}{(\text{tensile strain})}$ (N/m^2)

$$= \dfrac{\dfrac{F}{A}}{\dfrac{\Delta L}{L}} = \dfrac{FL}{A\Delta L}$$

Volumetric elasticity

Volumetric stress: Δp (N/m^2)

p = pressure of the object causing the stress

Volumetric strain: $-\dfrac{\Delta V}{V}$

ΔV = change in volume as a result of stress
V = original volume of material

Bulk modulus: $B = \dfrac{(\text{volumetric stress})}{(\text{volumetric strain})}$ (N/m^2)

$$= -\dfrac{\Delta p}{\dfrac{\Delta V}{V}} = -V\dfrac{\Delta p}{V}$$

Shear elasticity

Shear Stress: $\dfrac{F}{A}$ (N/m^2)

F = exerted tangential force causing the stress
A = cross-sectional area of surface being sheared

Shear strain: $\dfrac{\Delta L}{L}$

ΔL = change in length as a result of shear
L = distance between sheared and original state

Shear modulus: $S = \dfrac{(\text{shear stress})}{(\text{shear strain})}$ (N/m^2)

$$= \dfrac{\dfrac{F}{A}}{\dfrac{\Delta L}{L}} = \dfrac{FL}{A\Delta L}$$

curve remains linear at the initial stages and then makes an abrupt turn, increasing rapidly. The modulus of the material can be determined by the slope of the curve only within portions of the curve that are linear. At points distal to the inflection point, an incremental elastic modulus must be calculated which implies a continually changing elastic modulus. For these materials, the elastic modulus must be determined according to the unique point of stress and strain under consideration.

2.5   Elastic Modulus

Elastic moduli are physical constants that describe the elastic properties of a solid. Values of elastic moduli for common physical and biological materials are given in Table 2.2. From Table 2.2, it should be obvious intuitively that steel would have a much higher elastic modulus than rubber, as it is considerably stronger and much harder to deform. These constants, in units of dyn/cm^2, are defined as the ratio of stress component to the corresponding type of strain component. One elastic modulus, given as the ratio of the tensile or longitudinal stress S_{xx} to the tensile strain ε_{xx} is the Young's modulus of the material, denoted by E:

$$E = \frac{S_{xx}}{\varepsilon_{xx}.} \tag{2.7}$$

TABLE 2.2. Values of Young's modulus for various materials.

Material	Young's modulus (dyn/cm^2)
Blood Vessel Components	
Collagen	$1 \times 10^8 – 1 \times 10^9$
Elastin	5×10^6
Smooth muscle	$1 \times 10^5 – 1 \times 10^7$
Biological Materials	
Bone	1×10^{11}
Tooth enamel	5×10^{11}
Tooth dentin	1×10^{11}
Hair	2×10^9
Skin	1×10^7
Textiles	
Polyethylene	3×10^9
Vulcanized rubber	5×10^5
Wood	1×10^{11}
Steel	2×10^{12}
Aluminum	7×10^{10}
Copper	11×10^{10}
Glass	7×10^{10}

It can also be expressed in terms of physical parameters unique to an elastic solid such as a wire or rod as

$$\text{Young's modulus} = Y = \frac{(\text{stress})}{(\text{strain})} = \frac{FL}{A}\Delta L,$$

where F is the force applied to the solid, A is the cross-sectional area of the solid, L is the original length of the solid, and ΔL is the fractional amount of deformation as a result of the applied force. Young's modulus was introduced in the previous section on Hooke's law as a result of tensile stress and strain. As mentioned in the section on stress (Sec. 2.2), there are three types of stresses or strains that thus possess different elastic moduli.

Another elastic modulus, defined as the shear modulus G, relates the shear stress S_{xy} to the angular strain ε_{xy} and is given by

$$G = \frac{S_{xy}}{\varepsilon_{xy}}.$$

In terms of physical parameters unique to the elastic solid, the shear modulus is equal to the expression of Young's modulus given by

$$\text{Shear modulus} = G = \frac{(\text{stress})}{(\text{strain})} = \frac{FL}{A}\Delta L. \tag{2.8}$$

However, it should be noted that the strain or $\Delta L/L$ is generally small and thus can be simplified by a small-angle approximation

$$\frac{\Delta L}{L} \approx \phi,$$

where ϕ is the angle by which the elastic solid is displaced by the shearing stress [Fig. 2.2(D)].

The elastic modulus, due to volumetric compression, describes the elasticity of the volume of the elastic solid. This elastic modulus is termed the bulk modulus B and is given by

$$B = -V\frac{\Delta P}{\Delta V}, \tag{2.9}$$

where V is volume and P is pressure. The minus sign is used to negate the negative effects of volumetric change, making the bulk modulus positive. Because the volumetric units cancel, the bulk modulus has units of pressure or dyn/cm^2.

2.6   Elastic Limit

Returning to the descriptive example of the rubber band presented at the beginning of this chapter, it is a true statement that the application of physical forces on the rubber band does not ordinarily result in a permanent

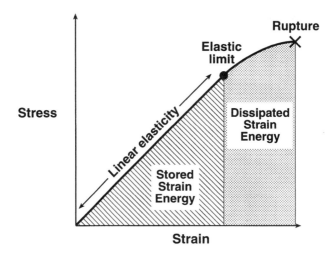

FIGURE 2.3. Stress–strain curve for a perfectly elastic object that proceeds linearly until a critical point, known as the elastic limit, is reached. The elastic limit typically signifies a permanent or irreversible deformation. Further stress exerted on the elastic material results in rupture, representing release of the stored strain energy.

deformation. However, it is obvious intuitively that, if a large enough force is applied, the rubber band will stretch irreversibly and eventually break. In this case, the rubber band exhibits linear elasticity, as shown by the linear diagonal relationship between the stress and strain in the curve displayed in Fig. 2.3, to an inflection point, where the curve becomes constant. The physical explanation for this inflection point is the *elastic limit* of the object. The elastic limit is the point of permanent deformation and represents the stored strain energy within an object. As the stress continues past the elastic limit, the elastic material eventually ruptures, whereby all the stored strain energy is released. The area contained under the curve from the elastic limit to the point of rupture is dissipated strain energy. The total strain energy characteristic of an elastic material is the sum of the stored strain energy and the dissipated strain energy.

The concept of elastic limit plays a critical role in cerebrovascular disease, particularly in the biophysical assessment of hemorrhagic stroke (Chap. 5). Consider the case of intracranial aneurysms that manifest as a ballooning or sacculations of a weakened region of blood vessels (Chap. 6). These sacculations can expand and possibly rupture, leading to similar consequences described for the previous mechanism of stroke. Although tissue metabolism may be a primary factor in the development and rupture of aneurysms, aneurysm rupture occurs when the intra-aneurysmal pressure exceeds the elastic modulus of the weakened region of the aneurysm. The elastic limit or point prior to aneurysm rupture, although well founded based on biophysical reasoning, is extremely difficult to diagnose or predict on a clinical basis.

The elasticity of blood vessels in health and disease (intracranial aneurysms) present unique cases of elastic volumes with different geometries. However, biophysical characterization of the elasticity of both a cylindrical blood vessel and a spherical aneurysm can be performed by analysis of inherent forces and the derivation of an expression for tension and stress based on these forces. The resultant derivations for stress are known as Laplace's law and will now be derived for both the elastic cylinder and sphere.

2.7   Laplace's Law

Laplace's law describes the relation between the circumferential stress and the radius for any curved elastic surface. Laplace's law, in effect, is based on Newton's third law and equates a force F_P produced by the transmural blood pressure over the cross-sectional area of the structure to a circumferential force (stress) S that compensates for the distension and is required to maintain equilibrium. Laplace's law has been applied to various organs and structures of the human body including the heart,[7-14] lungs,[15-20] stomach,[21] muscle,[22] nerves,[23-26] blood vessels,[27-33] eye,[34-36] diaphragm,[37,38] colon,[39] bowel,[40] ureter,[41] bladder,[42,43] and uterus.[44] The intrigue and attractiveness of applying Laplace's law to the organs and structures of the human body arise primarily from the geometric and physical similarities to elastic spheres and cylinders. In the majority of incidences, Laplace's law can be elegantly employed to model qualitative and, in some instances, quantitative aspects of organ function in normal and diseased states. Laplace's law is dependent strongly on the distribution of stress, which varies according to the geometry of the curved surface. Specific applications of Laplace's law to cerebrovascular diseases include the quantitation of stress in a normal blood vessel with cylindrical geometry and in an aneurysm with spherical geometry and thus require appropriate geometric transformations.

2.7.1   Laplace's Law for an Elastic Cylinder

In order to properly derive an expression for Laplace's law, regardless of the geometry, one must resolve the physical forces and resulting stresses acting on the elastic structure with respect to the geometry. Consider the diagram of the cylindrical vessel with an internal radius R_i, an external radius R_e, length L, and wall thickness h in Fig. 2.4. In static equilibrium, there are two forces exerted along the vessel wall in opposing directions. The first force results from a circumferential stress S_θ acting along the curved portion of the vessel and is defined mathematically by[45]

$$F_\theta = 2S_\theta Lh, \qquad (2.10)$$

while the second opposing force is the result of the intraluminal pressure P_i

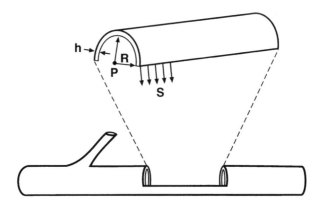

FIGURE 2.4. Graphical description of the forces acting on an elastic cylinder according to Laplace's law.

acting outward against an external pressure P_e in a radial direction:

$$F_p = P_i(2R_iL) - P_e(2R_eL). \qquad (2.11)$$

Since these are the only two forces acting on the cylinder, they may now be equated:

$$2S_\theta Lh = P_i(2R_iL) - P_e(2R_eL).$$

Solving for S_θ yields the equation for stress for an elastic cylinder:

$$S_\theta = \frac{P_iR_i}{h} - \frac{P_eR_e}{h}. \qquad (2.12)$$

This relation, given by Eq. (2.12), is known as Laplace's law for an elastic cylinder. For most applications, particularly when dealing with the small blood vessels of the cerebrovasculature, the vessel wall is assumed to be very thin, i.e., $h \ll R$, implying that $R_i \approx R_e$, and the stress now becomes

$$S_\theta = \frac{PR}{h}. \qquad (2.13)$$

At first glance, when considering Laplace's law for an elastic cylinder, one might be tempted to focus solely on the relationship between the stress and radius. However, the parameters P and h are just as important in describing the elasticity of an elastic cylinder and can be illustrated through the differential of the expression for stress:[46]

$$dS = d\frac{PR}{h} = \frac{R}{h}dP + \frac{P}{h}dR - \frac{PR}{h^2}dh.$$

As an elastic structure acquires wall thickness, i.e., $h < R$ as opposed to $h \ll R$, the circumferential stress drops off from a maximum value at the inner wall to a minimum value at the outer wall, the drop being greater the larger the ratio of diameters.[47]

An example illustrating the importance of elasticity in cerebrovascular disease is the changes in wall elasticity of an artery as a result of distension due to intravascular pressure. The normal artery contains the structural proteins elastin and collagen. Although collagen exhibits a larger elastic modulus in comparison to elastin, elastin contributed strongly to the distensibility and general state of elasticity of the artery. From experimental studies on the elastic properties of human iliac arteries, Roach and Burton[48] concluded that the resistance to stretch at low pressures was almost entirely due to elastin fibers, that at physiologic pressures it was due to both collagen and elastin fibers, and that at high pressures it was almost entirely due to collagen fibers.[49] These are believed to be the primary mechanisms for the loss of structural support in the development of intracranial aneurysms (Chap. 6). As an aneurysm develops and the artery wall begins to balloon or sacculate as a result of the distending pressure, the elastin fragments and plays a substantially reduced role in maintaining the structural elasticity of the diseased arterial wall. As a consequence, collagen assumes the majority of the exerted hemodynamic load and the aneurysm wall becomes less distensible.[50] This translates to a greater stress exerted on the aneurysm wall in response to the same strain, accelerating structural fatigue of the aneurysm wall and eventual rupture.

Laplace's law can also be expressed in terms of the elastic strain ε by

$$S = \varepsilon E, \tag{2.14}$$

where E is the elastic modulus of the elastic cylinder. The elastic modulus increases with increasing pressure, which means that the cylindrical vessel wall becomes less distensible with increasing intraluminal pressure.[51]

2.7.2 Laplace's Law for an Elastic Sphere

Proceeding in a manner similar to that outlined in the previous section (Laplace's law for an elastic cylinder), consider the free-body diagram shown in Fig. 2.5 applied to the cross-sectional geometry of a sphere with an internal radius R_i, an external radius R_e, and a wall thickness h. There are two forces acting on the spherical surface. One force, which acts toward the spherical surface, is that produced by the pressure, denoted by F_P. Pressure is a radial force that acts perpendicular to the vessel surface at every point along the cross-sectional area of the sphere. Thus, the force due to pressure, F_P, is simply the product of the pressure and the cross-sectional area of the sphere or

$$F_P = P\pi R^2, \tag{2.15}$$

where R is the radius of the sphere. F_P is countered by another force due to stress, F_S, acting in an opposite direction. F_S is the average stress experi-

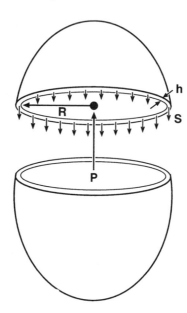

FIGURE 2.5. Graphical description of the forces acting on an elastic sphere according to Laplace's law.

enced by each point along the area of the spherical structure supporting this stress and is defined by

$$F_S = 2\pi RhS, \tag{2.16}$$

where h is the wall thickness of the spherical structure and S is the average stress.[52] Equating the forces presented by Eqs. (2.15) and (2.16) and solving for S yields

$$P\pi R^2 = 2\pi RhS,$$

or

$$S = \frac{PR}{2h}. \tag{2.17}$$

It should be noted that the above expression for Laplace's law (1) is valid only for thin-walled elastic structures, i.e., $h \ll R$ and (2) states a linear relationship between the circumferential stress and the radius of any curved surface, i.e., the stress required to maintain static equilibrium increases as the radius increases.

In addition, Laplace's law defines the stress for an elastic structure in static equilibrium, i.e., the structure remains stationary. In physiological applications, however, the structure is in constant dynamic motion due to the continual pulsatile flow of blood. Thus, although applications of Laplace's law exist in many aspects of medical science,[53] many limitations restrict the usefulness to qualitative observations rather than quantitative calculations.

2.8   Equations of Motion of Elastic Objects

Due primarily to its physical structure, an elastic object, when subjected to an external force or stress, will respond with vibrations or oscillatory fluctuations of energy as the motion of the material progresses toward equilibrium. The oscillatory fluctuations, as they propagate spatially along the volume of the elastic material, can be characterized physically as transverse waves. The dynamics and consequent physical interactions can be described for any elastic object through the derivation of an equation of motion. The equation of motion of an elastic object is derived typically by assessing the physical interactions involved in the source of the motion and translating these interactions into equation form. The basis for the equation of motion is Newton's second law or $F = ma$. The force responsible for the motion within an elastic object, which is pulsatile when considering biological vessels, is pressure or the force exerted over a given surface area. The motion or acceleration of the elastic object depends on the object geometry and the complexity of inherent assumptions. It should also be stated that the motion is similar in nature to vibrations rather than visible fluctuations. The explanation and derivation of equations of motion will be described for cylindrical and spherical geometries. The cylinder and sphere are geometric representations of physiological significance with the elastic cylinder representing a blood vessel (Chaps. 4 and 5) and the sphere representing an intracranial aneurysm (Chap. 6).

2.8.1   Equation of Motion of an Elastic Sphere

In biological applications, pulsatile blood flow is propagated through the human vasculature under a large systolic blood pressure. Thus, the primary objective of the equation of motion is to determine a relation between the periodic nature of the pressure and the resultant vibrational motion as a result of the pressure. Two approaches exist for the derivation of equations of motion of elastic objects and both of them will be applied to the elastic sphere.

Assume that we want to derive an equation of motion for an elastic sphere under the assumption that it is a thin shell. The parameter ξ, which is dependent on both position and time, represents the displacement at some later time from R to $R + \xi(R, t)$. In terms of the strain ε, the circumferential stress S of the expanded sphere is then given by

$$S = E\varepsilon = E\frac{2\pi(R + \xi) - 2\pi R}{2\pi R} = E\frac{\xi}{R}, \tag{2.18}$$

where E is the elastic modulus of the elastic sphere. S can also be expressed, according to Laplace's law for a spherical elastic object (see Sec. 2.7.2), as

$$S = \frac{PR}{2h}, \tag{2.19}$$

where P is the internal pressure.

An equation of motion describing the vibrational displacement of the elastic sphere can be derived by considering the forces acting on the sphere in static equilibrium. These include forces due to pressure (F_{pres}), circumferential stress (F_{stress}), and inertia (F_{iner}). F_{pres} is a distending force that acts in an outward direction, while the other forces, i.e., F_{iner} and F_{stress}, are compressive forces and act in a direction opposite to F_{pres}. The balance of forces written in equation form are

$$F_{pres} = F_{stress} + F_{iner}. \tag{2.20}$$

Since the motion of the elastic sphere in physiological applications is driven by a pulsatile pressure function, a more accurate characterization of the wall dynamics of the sphere can be obtained by expressing Eq. (2.20) as the balance of pressures (force per unit surface area) instead of forces:

$$P = P_{stress} + P_{iner}. \tag{2.21}$$

P represents the pulsatile driving force exerted on the wall of the sphere, which is identical to the internal pressure.

P_{stress} is the pressure or force exerted on the wall of the sphere by the circumferential stress and can be determined by setting Eq. (2.18) equal to Eq. (2.19) and solving for P:

$$\frac{PR}{2h} = E\frac{\xi}{R},$$
$$P_{stress} = 2hE\frac{\xi}{R^2}. \tag{2.22}$$

Equation (2.22) can be further simplified by

$$P_{stress} = B\xi, \tag{2.23}$$

where $B = 2hE/R^2$.

F_{iner} is the inertial force due to the acceleration of the sphere and, according to Newton's second law, is equal to

$$F_{iner} = ma, \tag{2.24}$$

where m is the object mass and a is the acceleration of the object. Thus, F_{iner} is represented by

$$F_{iner} = m_{wall}a = \rho_w \frac{R^2}{2} h \frac{d^2\xi}{dt^2}, \tag{2.25}$$

where ρ_w is the density of the wall of the sphere $(= 1.0\,g/cm^3)$ and $d^2\xi/dt^2$ is the acceleration of the wall in response to an external driving force. The pressure due to inertia, P_{iner}, is

$$P_{iner} = \rho_w h \frac{d^2\xi}{dt^2}. \tag{2.26}$$

Similar to the other components of pressure, P_{iner} can be simplified according to

$$P_{iner} = A \frac{d^2\xi}{dt^2}, \tag{2.27}$$

where $A = \rho_w h$.

Substituting the components of pressure given by Eqs. (2.26) and (2.22) into Eq. (2.21) yields the following differential equation:

$$P = \rho_w h \frac{d^2\xi}{dt^2} + 2hE \frac{\xi}{R^2}. \tag{2.28}$$

Equation (2.28), which is a linear, nonhomogeneous, second-order differential equation, can be rewritten as

$$P = A \frac{d^2\xi}{dt^2} + B\xi. \tag{2.29}$$

The pressure P can be represented as the driving force of the pulsatile blood flow and is stated mathematically as

$$P = F_{bp} \cos(\omega_{df} t), \tag{2.30}$$

where F_{bp} is the magnitude of the pulsatile force and ω_{df} is the frequency of the driving force. Equation (2.29) is then rewritten as

$$A \frac{d^2\xi}{dt^2} + B\xi = F_{bp} \cos(\omega_{df} t). \tag{2.31}$$

The general solution to the differential equation presented in Eq. (2.31), ξ, can be found in any elementary engineering textbook[54]:

$$\xi_P = \frac{F_{bp} \cos(\omega_{df} t - \theta)}{[(C - A\omega_{df}^2)^2 + \omega_{df}^2 B^2]^{1/2}}. \tag{2.32}$$

ω_{df} represents the frequency of the forced vibrations of the sphere.

Diverging just for a brief moment, a vibrating system with a given frequency exhibits a resonant frequency. The resonant frequency, which occurs when the frequency of the pulsatile driving force equals that of the vibrating object, has been considered as a viable mechanism behind the rupture of intracranial aneurysms (Chap. 6). A quantitative expression for the resonant frequency can be derived from the solution to the equation of motion for an elastic sphere.

The sphere is constantly vibrating at the natural damped frequency ω_{dnf}, the amplitude of which depends upon the initial conditions and decays exponentially with time. The resonant frequency is the frequency determined from the maximum of Eq. (2.32), which can be rewritten as

$$\left| \frac{\xi_P}{F_{bp}} \right| = \frac{1}{[(B - A\omega_{df^2})^2]^{1/2}} . \tag{2.33}$$

This can be simplified by squaring both sides, inverting, and differentiating with respect to ω_{df} to obtain the minimum of Eq. (2.33) or

$$\frac{d \left| \frac{F_{bp}}{\xi_p} \right|^2}{d\omega_{df}} = 2(B - A\omega_{df}^2)(-2A\omega_{df}) = 0.$$

Solving for ω_{df}, the resonant frequency of the elastic sphere due to wall displacement is

$$\omega_{df} = \omega_{res} = \left(\frac{C}{A} \right)^{1/2} . \tag{2.34}$$

One can also approach the same problem of vibrational motion in an elastic sphere using a more complex description of the physics and mathematics. We begin by making the same assumptions as in the previous derivation in that the elastic sphere is hollow, thin-walled, and remains stationary such that we are concerned only with motion in a radial direction. The dynamic equation for the radial displacement of an elastic sphere in spherical coordinates is given by the Navier–Stokes equation[55]:

$$\rho \left(\frac{\partial u_r}{\partial t} + u_r \frac{\partial u_r}{\partial r} + \frac{u_\theta}{r} \frac{\partial u_r}{\partial \theta} + \frac{u_\phi}{r \sin \theta} \frac{\partial u_r}{\partial \phi} - \frac{u_\theta^2 + u_\phi^2}{r} \right)$$

$$= - \left[\frac{\partial p}{\partial r} + \frac{1}{r^2} \frac{\partial}{\partial r} (r^2 T_{rr}) + \frac{1}{r \sin \theta} \frac{\partial}{\partial \theta} (T_{r\theta} \sin \theta) + \frac{1}{r \sin \theta} \frac{\partial T_{r\phi}}{\partial \phi} - \frac{T_{\theta\theta} - T_{\phi\phi}}{r} \right]$$

$$+ \rho g_r, \tag{2.35}$$

where ρ is the density of the fluid, u_r, u_θ, u_ϕ are the velocity components in the r, θ, ϕ directions, respectively, $\partial p / \partial r$ is the radial transmural pressure gradient, T_{rr}, $T_{\theta\theta}$, and $T_{\phi\phi}$ are the stress components in the radial, circumferential, and azimuthal directions, respectively, and g_r is the gravitational constant. The Navier–Stokes equations describe the three-dimensional motion of a viscous medium and will be described in more detail when applied to the physics of blood flow or hemodynamics (Chap. 4). Neglecting the contribution from the body-force term, i.e., ρg_r, the equation of motion describing the vibrational displacements of the elastic sphere in the radial direction is

$$\rho\left(\frac{\partial u_r}{\partial t} + u_r \frac{\partial u_r}{\partial r}\right) = -\frac{\partial p}{\partial r} + \frac{1}{r^2}\frac{\partial}{\partial r}(r^2 T_{rr}) - \frac{T_{\theta\theta} - T_{\phi\phi}}{r}. \tag{2.36}$$

The radial velocity u_r can be determined from the continuity equation which, given in spherical coordinates, is

$$\frac{1}{r^2}\frac{\partial}{\partial r}r^2 u_r + \frac{1}{r\sin\theta}\left(\frac{\partial}{\partial\theta}(u_\theta \sin\theta) + \frac{\partial u_\phi}{\partial\phi}\right) = 0. \tag{2.37}$$

Again, assuming that the elastic sphere is subject only to radial displacements, then Eq. (2.37) reduces to

$$\frac{1}{r^2}\frac{\partial}{\partial_r}(r^2 u_r) = 0.$$

Multiplying both sides by r^2 further simplifies the continuity equation to

$$\frac{\partial}{\partial r}(r^2 u_r) = 0.$$

Using elementary calculus, it can be seen that a solution for u_r is

$$u_r = \frac{C}{r^2}, \tag{2.38}$$

where C is a constant. In order to ensure continuity of the radial velocity, the following kinematic boundary condition is placed at the aneurysm wall, defined by radius R:

$$u_r|_R = \frac{dR}{dt}. \tag{2.39}$$

Thus, for any given radius u_r, the radial velocity is given as

$$u_r = \frac{dR}{dt}\frac{R^2}{r^2}. \tag{2.40}$$

The next step is to obtain an expression for the stresses T_{rr}, $T_{\theta\theta}$, $T_{\phi\phi}$. The general formula for the stress components is given by the stress tensor

$$T_{ij} = -p\delta_{ij} + \mu S_{ij}, \tag{2.41}$$

where T_{ij} represents the stress components, $-p$ is the transmural pressure, δ_{ij} is the Kronecker delta function, which equals 1 if $i = j$ and 0 if $i \neq j$, μ is the viscosity of the fluid, and S_{ij} is the rate of strain deformation. The spherical components of the rate of strain deformation are given by

$$S_{rr} = 2\frac{\partial u_r}{\partial r}, \tag{2.42a}$$

$$S_{\theta\theta} = 2\left(\frac{1}{r}\frac{\partial u_\theta}{\partial \theta} + \frac{u_r}{r}\right), \tag{2.42b}$$

$$S_{\phi\phi} = 2\left(\frac{1}{r\sin\theta}\frac{\partial u_\phi}{\partial \phi}\frac{u_r}{r} + \frac{u_\theta\cot\theta}{r}\right). \tag{2.42c}$$

Since only radial displacements are considered, Eqs. (2.42b) and (2.42c) can be simplified to

$$S_{\theta\theta} = S_{\phi\phi} = 2\frac{u_r}{r}. \tag{2.43}$$

From Eq. (2.40), the components for the rate of strain deformation are

$$S_{rr} = 2\frac{dR}{dt}\left(-2\frac{R^2}{r^3}\right) = -4\frac{dR}{dt}\frac{R^2}{r^3}, \tag{2.44}$$

$$S_{\theta\theta} = S_{\phi\phi} = 2\frac{u_r}{r} = 2\frac{dR}{dt}\frac{R^2}{r^2}\frac{1}{r} = 2\frac{dR}{dt}\frac{R^2}{r^3}. \tag{2.45}$$

All cross terms of the rate of strain deformation, i.e., $S_{r\theta}$, $S_{r\phi}$, $S_{\theta\phi}$, are zero. Using Eqs. (2.44) and (2.45), the stress components can now be calculated from Eq. (2.41):

$$T_{rr} = -p - 4\mu\frac{dR}{dt}\frac{R^2}{r^3}, \tag{2.46a}$$

$$T_{\theta\theta} = -p + 2\mu\frac{dR}{dt}\frac{R^2}{r^3}, \tag{2.46b}$$

$$T_{\phi\phi} = -p + 2\mu\frac{dR}{dt}\frac{R^2}{r^3}. \tag{2.46c}$$

The next step is to determine the internal pressure distribution within the elastic sphere. The internal pressure P responsible for wall displacement can be determined by imposing the following boundary condition regarding radial stress at the wall of the sphere,

$$-p = T_{rr}|_R = -p_R - \frac{4\mu}{R}\frac{dR}{dt}, \tag{2.47}$$

where p_R is the transmural pressure evaluated at R. The parameter p_R is determined from Eq. (2.36) by solving for the term $-\partial p/\partial r$ and evaluating the expression at the wall of the sphere, i.e., $r = R$:

$$\frac{\partial p}{\partial r} = \frac{1}{r^2}\frac{\partial}{\partial r}(r^2 T_{rr}) - \frac{T_{\theta\theta} - T_{\phi\phi}}{r} - \rho\left(\frac{\partial u_r}{\partial t} + u_r\frac{\partial u_r}{\partial r}\right).$$

Making the appropriate substitutions and performing the necessary mathematical operations yields

$$\frac{\partial p}{\partial r} = \frac{1}{r^2}\frac{\partial}{\partial r}\left[r^2\left(-p - 4\mu\frac{dR}{dt}\frac{R^2}{r^3}\right)\right]$$

$$-\left[\left(-p + 2\mu\frac{dR}{dt}\frac{R^2}{r^3}\right) - \left(-p + 2\mu\frac{dR}{dt}\frac{R^2}{r^3}\right)\right]r$$

$$-\rho\left[\frac{R^2}{r^2}\frac{d^2R}{dt^2} + \frac{dR}{dt}\frac{R^2}{r^2}\left(-2\frac{R^2}{r^3}\frac{dR}{dt}\right)\right],$$

$$\frac{\partial p}{\partial r} = \frac{1}{r^2}\frac{\partial}{\partial r}r^2\left(-p - 4\mu\frac{dR}{dt}\frac{R^2}{r^3}\right) \tag{2.48}$$

$$-\rho\left[\frac{R^2}{r^2}\frac{d^2R}{dt^2} + \frac{dR}{dt}\frac{R^2}{r^2}\left(-2\frac{R^2}{r^3}\frac{dR}{dt}\right)\right],$$

$$\frac{\partial p}{\partial r} = -\rho\left[\frac{R^2}{r^2}\frac{d^2R}{dt^2} + \frac{dR}{dt}\frac{R^2}{r^2}\left(-2\frac{R^2}{r^3}\frac{dR}{dt}\right)\right],$$

$$\frac{\partial p}{\partial r} = -\rho\left[\frac{R^2}{r^2}\frac{d^2R}{dt^2} + \left(\frac{dR}{dt}\right)^2\frac{R^4}{r^5}(-2)\right].$$

Integrating and evaluating at $r = R$ results in:

$$p_R = -\rho\left[-R\frac{d^2R}{dt^2} + \left(\frac{dR}{dt}\right)^2\left(\frac{-1}{4}\right)(-2)\right]. \tag{2.49}$$

Substituting Eq. (2.49) into Eq. (2.47):

$$p = \rho\left[R\frac{d^2R}{dt^2} - \left(\frac{dR}{dt}\right)^2\frac{1}{2}\right] + \frac{4\mu}{R}\frac{dR}{dt}. \tag{2.50}$$

As can be seen, the pressure-induced stresses are nonlinearly related to the radius of the sphere as well as the time derivatives of the radius.

2.8.2 Equation of Motion of an Elastic Cylinder

The equation of motion for an elastic cylinder can be derived from the Navier–Stokes equation as was done in the previous example for the elastic sphere. The details of the derivation will be omitted but the equations of motion according to the cylindrical components r, ϕ, and z are[56]

$$\rho\frac{\partial^2 u_r}{\partial t^2} = -\frac{\partial\Omega}{\partial r} + \mu_0\left(\frac{\partial^2 u_r}{\partial r^2} + \frac{1}{r}\frac{\partial u_r}{\partial r} - \frac{u_r}{r^2} + \frac{\partial^2 u_r}{\partial z^2} + \frac{1}{r^2}\frac{\partial^2 u_p}{\partial\phi^2} - \frac{2}{r^2}\frac{\partial u_\phi}{\partial\phi}\right),$$

$$\rho\frac{\partial^2 u_\phi}{\partial t^2} = -\frac{\partial\Omega}{\partial\phi} + \mu_0\left(\frac{\partial^2 u_\phi}{\partial r^2} + \frac{1}{r^2}\frac{\partial^2 u_\phi}{\partial\phi^2} + \frac{\partial^2 u_\phi}{\partial z^2} + \frac{1}{r}\frac{\partial u_\phi}{\partial r} + \frac{2}{r^2}\frac{\partial u_r}{\partial\phi} - \frac{u_\phi}{r^2}\right),$$

$$\rho\frac{\partial^2 u_z}{\partial t^2} = -\frac{\partial\Omega}{\partial z} + \mu_0\left(\frac{\partial^2 u_z}{\partial r^2} + \frac{1}{r^2}\frac{\partial^2 u_z}{\partial\phi^2} + \frac{\partial^2 u_z}{\partial z^2} + \frac{1}{r}\frac{\partial u_z}{\partial r}\right),$$

where ρ is the density of the vessel wall ($\approx$ density of blood), Ω is an expression for isotropic pressure $[= (S_{rr} + S_{\phi\phi} + S_{zz})/3]$, and μ_o is the elastic shear modulus. In most applications involving the computational simulation of fluid flow, the equation for ϕ is omitted since it is assumed typically that there is no rotational component to the fluid flow. In mathematical calculations and computational simulations of fluid flow, these equations of motion for the elastic object are linked dynamically to the motion of the pulsatile blood flow through boundary conditions.

2.9   Viscoelasticity

Up to this point, we have been concerned primarily with perfectly elastic objects or objects that deform as a result of an applied stress but return to original form once the force is removed. From a different perspective, the perfectly elastic object stores energy as a result of the deformation but is utilized, following removal of the stress, for transformation to the original form. These elastic materials are considered ideal and are not truly representative of the majority of elastic materials including biological materials. Biological materials also store energy during the application of stress, but dissipate energy in a timely manner, not immediately upon removal of the exerted stress. In other words, the strain exhibited by such materials becomes a time dependent entity. The time-dependence of strain is due to the viscous nature of biological materials, and the elasticity is described as viscoelasticity. The more viscous the material becomes, the longer the time required for the deformation to reach its limit.[57] For a comparison between the two types of elasticity, the stress of a perfectly elastic object can be described according to Hooke's law or

$$S = E\varepsilon.$$

For a viscoelastic material, the Young's modulus becomes a function of frequency and hence a complex entity according to[58]

$$E_c(\omega) = \frac{S(\omega)}{\varepsilon(\omega)} = |E_c(\omega)|e^{i\phi(\omega)}, \quad |E_c(\omega)| = \frac{L_o}{S}\left|\frac{F(\omega)}{\Delta L(\omega)}\right|. \tag{2.51}$$

Linear Elasticity

Viscoelasticity

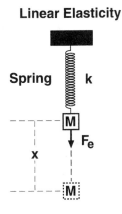

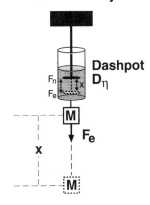

Spring **k**

Dashpot **D$_\eta$**

x

x

$$F_e = - kx$$

$$F_e = - D_\eta \frac{dx}{dt}$$

Stress = Elastic Modulus x Strain

Stress = Viscosity x Strain Rate

F$_e$ = external force
k   = spring constant
D$_\eta$ = dashpot viscosity

FIGURE 2.6. Analog models describing the physical interactions of elastic and visco-elastic materials.

In terms of the viscosity and strain rate, the stress of a viscoelastic object can be described by the expression

$$S = \eta \frac{d\varepsilon}{dt}, \qquad (2.52)$$

where η is the viscosity. Viscosity, an inherent property of all elastic materials to some extent, is an internal resistive force against applied stresses and the resultant deformation. Although viscosity is usually considered as a descriptive physical parameter for fluids and will be discussed in detail in Chap. 4, it is also important in characterizing the behavior of viscoelastic solids. The differences between these two types of elasticity can be visualized by implementing physical models of elasticity, shown in Fig. 2.6. The physical model for a linearly elastic solid or a solid that obeys Hooke's law is characterized by a force acting in a direction opposite to the spring's point of attachment. The force F causes a displacement x, dependent primarily on the spring constant. The viscoelastic model is arranged similar to the previous model with the exception that, instead of a spring, a dashpot is used. A dashpot consists of a piston encased within a cylinder filled with viscous fluid. As a force is exerted on the cylinder, the motion of the cylinder is damped to the viscous retarding force acting against the piston.

Application of viscoelastic approximations to the blood vessel can be made by incorporating the time dependence of strain into Laplace's law for an elastic cylinder:

$$S = E\varepsilon + R\frac{d\varepsilon}{dt}.$$

R is a proportionality constant between the pressure and rate of strain that is dependent on wall viscosity and geometry.[59] Wall thickness, particularly as it applies to blood vessels, is not constant and is dependent on stress. For a detailed description of the mathematics and physics of viscoelasticity, the reader is referred to the textbooks listed in Refs.[60-62]

2.10   Summary

In this chapter, the fundamental principles of elasticity were reviewed. The properties responsible for the physical behavior of elasticity are stress, strain, and elastic modulus. Since the blood vessels of the human body are elastic and subjected continually to stresses from the pulsatile ejection of blood from the heart, it is the capacity of the vessels to withstand these pulsatile stresses that, in most cases, leads to the onset and development of cerebrovascular disease. Physical relations such as Laplace's law allow one to both qualitatively and quantitatively represent the stability of these elastic vessels through elementary analysis of the forces presented by the hemo-dynamics. The type of elasticity characteristic of many materials, including biological materials, is viscoelasticity.

2.11   References

1. L.H. Greenberg, *Physics for Biology and Pre-Med Students* (Saunders, Phila-delphia, 1975), p. 325.
2. W.R. Milnor, *Hemodynamics*, 2nd ed. (Williams & Wilkins; Baltimore, 1989), p. 58.
3. D.J. Patel and R.N. Vaishnav, *Basic Hemodynamics and Its Role in Disease Processes* (University Park, Baltimore, 1980), p. 68.
4. E.O. Attinger, *Pulsatile Blood Flow* (McGraw-Hill, New York, 1964), p. 58.
5. W.W. Nichols and M.F. O'Rourke, *McDonald's Blood Flow in Arteries: Theo-retic, Experimental and Clinical Principles* (Lea & Febiger, Philadelphia, 1990), p. 81.
6. W.R. Milnor, *Hemodynamics*, 2nd ed. (Williams & Wilkins, Baltimore, 1989), p. 60.
7. M. Sugawara, K. Tamiya, and K. Nakano, "Regional work of the ventricle: wall tension–area relation," Heart Vessels 1985, 133–144.
8. R.R. Martin, and H. Haines, "Application of Laplace's law to mammalian hearts," Comp. Biochem. Physiol. **34**, 959–962 (1970).

9. D.G. Gordon, "The physics of left ventricular ejection and its implications for muscle mechanics," Eur. J. Cardiol. **4** (Suppl), 87–95 (1976).

10. R.J. Tallarida, B.F. Rusy, and M.H. Loughnane, "Left ventricular wall acceleration and the law of Laplace," Cardiovasc. Res. **4**, 217–223 (1970).

11. T.F. Moriarty, "The law of Laplace. Its limitations as a relation for diastolic pressure, volume, or wall stress of the left ventricle," Circ. Res. **46**, 321–331 (1980).

12. B.A. Carabello, "The relationship of left ventricular geometry and hypertrophy to left ventricular function in valvular heart disease," J. Heart Valve Dis. **4** (Suppl 2): S132–138, discussion S138–139 (1995).

13. G. Bittar, "Hemodynamic stress in hypertrophic cardiomyopathy, blood pressure changes: the age factor/the law of Laplace" [letter; comment], Circulation **84**, 1877 (1991).

14. J.K. Li, "Comparative cardiac mechanics: Laplace's law," J. Theor. Biol. **118**, 339–343 (1986).

15. S.B. Hall, M.S. Bermel, Y.T. Ko, H.J. Palmer, G. Enhorning, and R.H. Notter, "Approximations in the measurement of surface tension on the oscillating bubble surfactometer," J. Appl. Physiol. **75**, 468–477 (1993).

16. J.H. Comroe, Jr., "Premature science and immature lungs. Part III. The attack on immature lungs," Am. Rev. Respir. Dis. **116**, 497–518 (1977).

17. G. Enhorning, "Pulsating bubble technique for evaluating pulmonary surfactant," J. Appl. Physiol. **43**, 198–203 (1977).

18. R. Reifenrath, "The significance of alveolar geometry and surface tension in the respiratory mechanics of the lung," Resp. Physiol. **24**, 115–137 (1975).

19. J.F. Kurfees, "The legacy of Laplace," J. Kentucky Med. Assoc. **78**, 197–202 (1980).

20. R. Ortega, "A simple device to demonstrate the law of Laplace" [letter], Anesth. Analg. **75**, 307 (1992).

21. D.F. Stubbs, "Models of gastric emptying," Gut **18**, 202–207 (1977).

22. O.M. Sejersted and A.R. Hargens, "Intramuscular pressures for monitoring different tasks and muscle conditions," Adv. Exp. Med. Biol. **384**, 339–350 (1995).

23. R.E. Strain and W.H. Olson, "Selective damage of large diameter peripheral nerve fibers by compression: an application of Laplace's law," Exp. Neurol. **47**, 68–80 (1975).

24. G. Margolis, (Letter) "Laplace's law" (Letter), J. Am. Med. Assoc. **231**, 811 (1975).

25. R.E. Strain, Jr. and W.H. Olson, "Tinel's sign and Laplace's law." (Letter), N. Engl. J. Med. **291**, 801–802 (1974).

26. R.E. Strain, Jr., R.S. Strain, and W.H. Olson, "Laplace's law, pressure neuropathy, tabes dorsalis, tic douloureux, facial spasm" (Letter), J. Am. Med. Assoc. **229**, 1864 (1974).

27. R.S. Tracy, T.J. Heigle, and M. Velez-Duran, "Evidence for the failure of the Laplace law as a sole explanation for wall thickening of arteries in hypertensive and aging normotensive kidneys," Arch. Pathol. Lab. Med. **113**, 342–349 (1989).

28. S.H. White, "Small phospholipid vesicles: internal pressure, surface tension, and surface free energy," Proc. Nat. Acad. Sci. **77**, 4048–4050 (1980).

29. M. Dujovny, N. Wackenhut, N. Kossovsky, L. Leff, C. Gomez, and D. Nelson,

"Biomechanics of vascular occlusion in neurosurgery," Acta Neurol. Latino-americana **26**, 123–127 (1980).

30. P. Borgstrom and P.O. Grande, "Myogenic microvascular responses to change of transmural pressure. A mathematical approach," Acta Physiol. Scand. **106**, 411–423 (1979).

31. E.L. Schiffrin, "Reactivity of small blood vessels in hypertension: relation with structural changes. State of the art lecture," Hypertension **19** (2 Suppl), II1–II9 (1992).

32. N. Iida, "Physical properties of resistance vessel wall in peripheral blood flow regulation—I. Mathematical model," J. Biomech. **22**, 109–117 (1989).

33. B. Folkow, "Structure and function of the arteries in hypertension," Am. Heart J. **114**, 938–948 (1987).

34. M. Cahane and E. Bartov, "Axial length and scleral thickness effect on susceptibility to glaucomatous damage: a theoretical model implementing Laplace's law," Ophthal. Res. **24**, 280–284 (1992).

35. K. Szczudlowski, "Glaucoma hypothesis: application of the law of Laplace," Med. Hypoth. **5**, 481–486 (1979).

36. M. Davanger, "Descemetocele and the law of Laplace," Acta Ophthal. **49**, 715–718 (1971).

37. M. Paiva, S. Verbanck, M. Estenne, B. Poncelet, C. Segebarth, and P.T. Macklem, "Mechanical implications of in vivo human diaphragm shape," J. Appl. Physiol. **72**, 1407–1412 (1992).

38. W.A. Whitelaw, L.E. Hajdo, and J.A. Wallace, "Relationship among pressure, tension, and shape of the diaphragm," J. Appl. Physiol. **55**, 1899–1905 (1983).

39. D. Light, "Laplace's law" (Letter), Am. Fam. Physician **33**, 46 (1986).

40. F. Hinman, Jr., "Pascal, Laplace and a length of bowel," J. D. Urologie **95**, 11–14 (1989).

41. P. Biancani, M. Hausman, and R.M. Weiss, "Effect of obstruction on ureteral circumferential force-length relations," Am. J. Physiol. **243**, F204–F210 (1982).

42. F. Ronchi, V.E. Pricolo, L. Divieti, M. Palmi, L. Brigatti, and G.M. Clement, "Experimental study on bladder wall's strain in vesical function," Urol. Res. **10**, 285–291 (1982).

43. G.K. Stillwell, "The Law of Laplace. Some clinical applications," Mayo Clin. Proc. **48**, 863–869 (1973).

44. R.J. Noveroske, "Two applications of the Law of LaPlace to obstetrics," J. Indiana State Med. Assoc. **72**, 416–417 (1979).

45. D.J. Patel and R.N. Vaishnav, *Basic Hemodynamics and Its Role in Disease Processes* (University Park, Baltimore, 1980), pp. 74–76.

46. S.C. Ling, H.B. Atabek, W.G. Letzing, and D.J. Patel, "Nonlinear analysis of aortic flow in living dogs," Circ. Res. **33**, 198–212 (1973).

47. A.V. Wolf, "Demonstration concerning pressure–tension relations in various organs," Science **115**, 243–244 (1952).

48. M.R. Roach and A.C. Burton, "Reason for the shape of the distensibility curves of arteries," Can. J. Biochem. Physiol. **35**, 681–690 (1957).

49. D.J. Patel and R.N. Vaishnav, *Basic Hemodynamics and Its Role in Disease Processes Baltimore*, (University Park, 1980), p. 183.

50. S. Scott, G.G. Ferguson, M.R. Roach. "Comparison, of the elastic properties of human intracranial arteries and aneurysm," Can. J. Physiol. Pharmacol. **50**, 328–332 (1972).

51. G.L. Papageorgiou and N.B. Jones, "Physical modelling of the arterial wall. Part 1: Testing of tubes of various materials," J. Biomed. Eng. **9**, 153–156 (1987).

52. D.J. Patel and R.N. Vaishnav, *Basic Hemodynamics and Its Role in Disease Processes* (University Park, Baltimore, 1980), p. 79.

53. G.K. Stillwell, "The law of Laplace: Some clinical applications," Mayo Clin. Proc. **48**, 863–869 (1973).

54. A.D. Nashif, D.I.G. Jones, and J.P. Henderson, *Vibration Damping* (Wiley, New York, 1985), pp. 122–123.

55. S. Middleman, *Fundamentals of Polymer Processing* (McGraw-Hill, New York, 1977), pp. 38–40.

56. A. Noordergraaf, *Circulatory System Dynamics* (Academic, New York, 1978) p. 123.

57. D.E. Strandness and D.S. Sumner, *Hemodynamics for Surgeons* (Grune & Stratton, New York, 1975), p. 12.

58. N. Westerhof and A. Noordergraaf, "Arterial viscoelasticity: A generalized model. Effect of input impedance and wave travel in the systematic tree," J. Biomech. **3**, 357–379 (1970).

59. L.H. Peterson, "Vessel wall stress–strain relationship," in *Pulsatile Blood Flow*, edited by E.O. Attinger (McGraw-Hill, New York, 1964), Chap. 15, p. 266.

60. A.E.H. Love, *A Treatise on the Mathematical Theory of Elasticity* (Cambridge University Press, Cambridge, 1927).

61. R.V. Southwell, *An Introduction to the Theory of Elasticity for Engineers and Physicists* (Oxford University Press, London, 1941).

62. J.A. Hudson, *The Excitation and Propagation of Elastic Waves* (Cambridge University Press, Cambridge, 1980).

2.12 Problems

2.1. What is the significance of the minus sign in Poisson's ratio σ?

2.2. Express Laplace's law for an elastic sphere in terms of its volume.

2.3. Laplace's law is based on which physical principle?

2.4. Given an elastic sphere with an intra-aneurysmal pressure of 150 mm Hg and a wall thickness $h = R/20$. What is the circumferential stress experienced by the wall of the elastic sphere?

2.5. How is a viscous material characterized in terms of its physical properties?

2.6. Consider the process of inflating a balloon of radius R. Derive an expression for the change in balloon radius with respect to time.

2.7. Express the complex elastic modulus, given by Eq. (2.52), in terms of trigonometric functions.

3
Circulation: Structure and Physiology

3.1 Introduction

Beginning at the early stages of embryonic development until death, the circulatory system is constantly circulating blood through the human body, assuring that all metabolic and biochemical processes required to sustain human life are operating at peak performance. Over an average life span of 80 years, the heart will beat about 3×10^9 times, at a little over one beat per second.[1] Continuing with the calculations for an average lifetime of 80 years, the heart will pump 2.4×10^8 liters (L) of blood over a distance of approximately 60 000 miles (96 500 km) per cardiac cycle, all at a considerably inefficient (<10%) capacity. In fact, if assigned the task, one would be hard-pressed to mechanically construct an accurate working model of the heart, even given state of the art materials, computational resources, and engineering techniques. It is in this chapter that the reader will begin to appreciate the number, magnitude, and complexity of the anatomical arrangement of the circulatory system and the corresponding physiological processes as they pertain to cerebrovascular diseases.

The circulatory system consists of a vast array of different types of blood vessels, arranged in a complex, intricate, yet efficient design to allow optimal permeation of blood to every point within the human body (Fig. 3.1). Although its mechanisms and consequences are diverse, the primary purpose of the circulation is to deliver oxygen and nutrients to all of the organs and tissues and to remove cellular metabolic waste products within the body. The brain of a normal adult typically maintains a blood flow of 50–55 mL per 100 g/min, corresponding to a volumetric blood flow rate of 800–1000 mL/min, and can adjust to both acute and chronic physiological changes such as increases or decreases in blood pressure to ensure a relatively constant flow. However, when the flow is compromised to any degree, the brain tissue is deprived of its vital oxygen and nutrients resources and the proper function of the brain is threatened. Cerebral blood flow dynamics is affected adversely during the course of cerebrovascular disease and plays a particularly important role in the outcome of a patient afflicted with cerebrovascular disease.

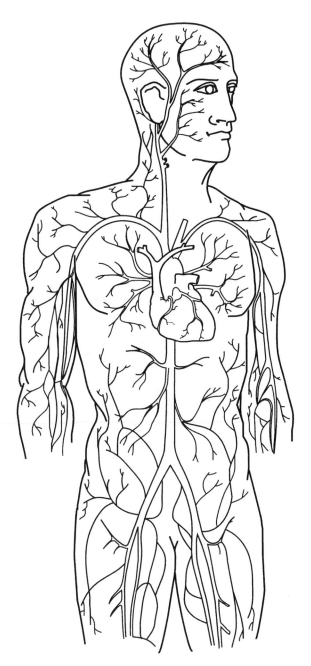

FIGURE 3.1. The arrangement of blood vessels comprising the human circulatory system.

This chapter is designed to introduce the reader to the elementary bio-physical and physiological principles and phenomena encountered in the circulatory system. Since cerebrovascular disease involves the blood vessels of the brain, the mechanics and dynamics of the circulation and resultant interactions with the blood vessels are instrumental in understanding the course of events involved in the initiation and development of these diseases. More specifically, this chapter addresses (1) the anatomy and biophysical function of the heart; (2) the identity and structure of the vascular compo-nents comprising the circulatory network; (3) the principles and experimental measurements of the biophysics and bioenergetics of the circulatory system; (4) the development of theoretical and *in vitro* models of the circulation; (5) the anatomy and physiology of the blood vessels comprising the cerebro-vasculature; and (6) the methods for obtaining experimental measurements of cerebral blood flow.

3.2   Physics of the Heart

The heart is an elastic, muscular double pump whose rhythmic contractions provide the force needed to circulate the blood throughout the network of blood vessels composing the circulatory system. Anatomically, the heart is composed of four chambers: two smaller chambers (right atrium and left atrium) contract systematically to eject blood into the two larger chambers (right ventricle and left ventricle). The right side propels blood to the lungs (for oxygen replenishment) and the left side propels it into the systemic circulation. The cycle of blood from the heart and through the circulation, illustrated in Fig. 3.2, will now be described.

The right atrium receives the systemic venous return through the supe-rior and inferior venae cavae (the largest veins in the body). Blood passes through a valve (the tricuspid valve) to enter the right ventricle. The right ventricle drives the blood from the heart to the pulmonary circula-tion under low pressure. Blood is pumped through another valve (the pulmonary valve) to enter the pulmonary artery on the way to both lungs. Oxygenated blood returns to the heart and enters the left atrium, then through the mitral valve into the left ventricle. This is the largest and most muscular of the four chambers and is responsible for pumping the blood out of the heart, through the aortic valve and the aortic arch, and around the rest of the body. The function of the left ventricle is to pump blood into the systemic circulation under high pressures ranging from 70 to 140 mm Hg.

The different parts of the heart normally beat in an orderly sequence: contraction of the atria (atrial systole) is followed by contraction of the ventricles (ventricular systole), and during diastole (rest) all four heart cham-bers are relaxed. The heartbeat originates in a specialized cardiac electrical conduction system and spreads via this system to all parts of the cardiac

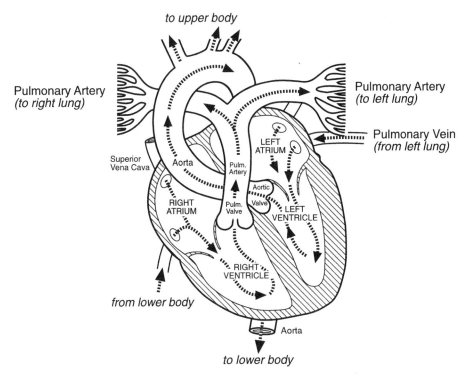

to upper body

Pulmonary Artery
(to right lung)

Pulmonary Artery
(to left lung)

Pulmonary Vein
(from left lung)

LEFT
ATRIUM

Superior
Vena Cava

Aorta

Pulm.
Artery

Aortic
Valve

RIGHT
ATRIUM

Pulm.
Valve

LEFT
VENTRICLE

RIGHT
VENTRICLE

from lower body

Aorta

to lower body

FIGURE 3.2. Schematic diagram of the circulation through the heart.

muscle. This triggers a wave of contraction that spreads, producing sequential changes in pressures and flows in the heart chambers and blood vessels. Systolic pressure refers to the peak pressure reached during systole, and diastolic pressure refers to the lowest pressure during diastole.

In late diastole, the mitral and tricuspid valves are open and the aortic and pulmonary valves are closed. Blood flows into the heart throughout diastole, filling both the atria and ventricles. The rate of filling declines as the ventricles become distended. The pressure in the ventricles remains low. Contraction of the atria occurs, propelling some additional blood into the ventricles. At the start of ventricular systole, the mitral and tricuspid valves close. Subsequent changes in pressure and volume of blood within the ventricles are shown graphically in Fig. 3.3. Intraventricular pressure rises sharply as the myocardium presses on the blood in the ventricle. This period (called isovolumetric ventricular contraction) lasts until the pressures in the left and right ventricles exceed those in the aorta (80 mm Hg) and the pulmonary artery (10 mm Hg), and then the aortic and pulmonary valves open. As these open, the phase of ventricular ejection begins. This is rapid at first, and then slows down. Peak left ventricular pressure reaches about 120 mm Hg, and peak right ventricular pressure is 25 mm Hg or less. The

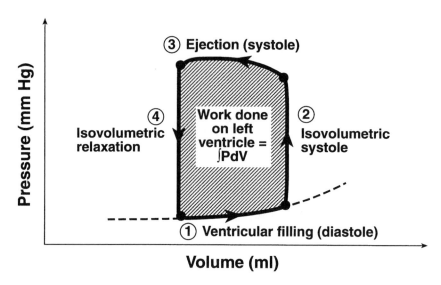

FIGURE 3.3. The work diagram depicting the pressure vs volume of the left ventricle during a cardiac cycle. The area within the curve represents the work done by the blood on the left ventricle during a left ventricular contraction.

amount of blood ejected by each ventricle per stroke at rest is 70–90 mL (the stroke volume). Once the ventricular muscle is fully contracted, the already falling ventricular pressures drop more rapidly. The aortic and pulmonary valves close, and pressure continues to drop rapidly during the period of isovolumetric ventricular relaxation. This ends when the ventricular pressure falls below atrial pressure and the mitral and tricuspid valves open, allowing the ventricles to fill. At a heart rate of 75 beats per minute, systole lasts about 0.27 s and diastole lasts about 0.53 s, with a total duration for each cardiac cycle of 0.80 s.

The output of the heart per unit time is the *cardiac output*. At rest, it averages about 5.5 L/min (80 mL × 69 beats/min). Variations in cardiac output can be produced by changes in cardiac rate or stroke volume. The cardiac rate is controlled primarily by the cardiac innervation. The stroke volume is also determined in part by neural input but varies with the length of the cardiac muscle fibers (i.e., its *preloading* or the amount these are allowed to stretch before contracting). The latter is dependent on many factors including the end-diastolic volume of blood in the heart (i.e., the amount of blood that can be accommodated by the heart before it is pumped out), the pressure within the thoracic cavity, the heart muscle stiffness, which may be altered in diseased states, and the amount of blood returning to the heart from the periphery. The strength of cardiac contraction is also determined by the resistance against which the ventricles pump blood (the *afterload*). This resistance is low in the pulmonary artery, but high in the aorta, being proportional to the resistance to flow through the aortic valve and the systemic blood pressure.

Blood ejected from the left ventricle propagates through a multitude of connected arteries that are initially large in size but systematically taper until the smallest arteries are reached (arterioles). From the arterioles, the blood then permeates through a wide expanse or bed of extremely small capillary vessels, which slow blood flow significantly to allow the blood to provide oxygen and nutrients to the surrounding tissue and organs while removing cellular waste. The capillaries converge into small veins (venules) that drain the capillary bed under small pressure. The assembly of veins is the opposite of arteries in that they originate from the capillary bed as small vessels and continually enlarge until the veins reach their largest size before returning deoxygenated blood back to the heart. Upon reaching the heart the blood first pools into the right atrium and enters the right ventricle, where it is pumped via the pulmonary artery to the lungs for oxygenation. Once the blood is oxygenated, it reenters the heart via the left atrium before making another trip through the systemic vasculature.

3.3   Circulatory Network: General Considerations

Blood vessels are a closed system of conduits that carry blood from the heart to the tissues and back to the heart. The characteristics of the various types of blood vessels are shown in Table 3.1. Blood flows through the vessels mainly because of the forward motion imparted to it by the pumping of the heart. In the systemic circulation this is aided by diastolic recoil of walls of arteries, compression of the veins by surrounding muscles, and the negative pressure in the thorax during inspiration. The resistance to flow depends to a minor degree on the viscosity of blood but mostly on the diameter of the vessels, primarily the arterioles (the resistance vessels). The systemic circulation is made up of numerous different circuits in parallel

TABLE 3.1. Biophysical parameters of the major vessels of the human body.

Vessel	Diameter (mm)	Length (cm)	Wall thickness (mm)	Wall tension (dyn/cm)	Internal pressure (mm Hg)
Large artery	8.0	20.0	1.0	90 000	97
Medium artery	4.0	15.0	0.8	60 000	90
Small artery	2.0	10.0	0.5	25 000	75
Arteriole	0.3	0.2	0.02	1 200	60
Capillary	0.008	0.075	0.001	16	30
Venule	0.02	0.2	0.002	26	20
Small vein	3.0	10.0	0.02	200	18
Medium vein	5.0	15.0	0.5	400	15
Large vein	15.0	20.0	0.8	9 750	10

(including that to the brain), an arrangement that permits wide variations in regional blood flow without changing total systemic flow.

3.3.1 Blood Vessel Types and Structure

The circulatory system is composed primarily of three types of blood vessels: arteries, veins, and capillaries. Arteries transport oxygen-rich blood from the heart under high pressure to beds of capillary vessels embedded in tissues and organs of the human body. The size of the various arteries located between the heart and capillary bed ranges from the large-sized arteries (2.5 cm) originating from the heart that gradually decrease to the size of arterioles (0.3 mm) feeding into the vascular bed of capillary vessels. The role of the veins is the exact opposite of arteries in that they transport oxygen-poor blood away from the capillary bed under low pressure and return to the heart.

The walls of the aorta (the largest artery in the body) and other large arteries contain a relatively large amount of elastic tissue. They stretch during systole and recoil during diastole. The walls of the arterioles contain less elastic tissue but much more smooth muscle. These vessels are the major site of the resistance to blood flow, and small changes in their caliber cause large changes in the total peripheral resistance. The arterioles divide into smaller vessels that feed into capillaries. These are narrow enough to allow circulating red blood cells to squeeze through in "single file." The walls of capillaries are made up of a single layer of endothelial (or lining) cells of 1 μm thickness. In the brain, the junctions between endothelial cells are tight, and the small gaps present only allow the passage of small molecules (<10 nm in diameter) from the blood to the brain cells. The total area of all the capillary walls in the body exceeds 6300 m^2 in the adult. The capillaries turn into venules with walls that are only slightly thicker than those of the capillaries. The walls of the veins are also thin and can be easily distended. They contain relatively little muscle, but considerable constriction can take place as a result of chemical and neural stimuli.

Arteries have relatively thicker walls and smaller lumina than veins. The basic structure of an artery consists of (1) a tunica intima, composed of the inner endothelium (usually one or two cell layers thick), a subendothelial layer of connective tissue, and an internal elastic lamina; (2) a tunica media, composed predominantly of circular smooth muscle fibers and interspersed fine elastic fibers; and (3) a tunica adventitia, composed of connective tissue. When arteries acquire about 25 or more layers of smooth muscle in the tunica media, they are referred to as medium-sized arteries. Elastic fibers become more numerous, but are still present as thin fibers and networks. The smallest arterioles have a thin internal elastic membrane and one layer of muscle in the media. The basic structure of a vein consists of (1) a tunica intima, composed of endothelium and a very thin layer of fine collagenous and elastic fibers that blend with the connective tissue of the media; (2) a tunica media consisting of a thin layer of circularly arranged smooth muscle fibers loosely embedded in connective tissue; and (3) a tunica adventitia consisting of a wide layer of connective tissue.

3.3.2 Mechanical Properties of Vessels

With the onset of each heartbeat, all blood vessels but particularly the aorta and arteries, which serve as the single mode of blood transport from the heart, are subjected continually to oscillatory hemodynamic forces. The ability of the blood vessel to withstand hemodynamic stresses and the resultant deformation is due primarily to its elasticity. Although the vessel elasticity is generally capable of bearing the stresses and resultant load of the hemodynamic forces, the deformation of the vessel wall represents a cumulative effect of all stresses exerted on the wall over time. The degree of vessel wall deformation is dependent on a host of factors including age, vessel tortuosity, patient history of hypertension and obesity among others, and secondary health risk factors such as smoking and alcohol consumption. A combination of these factors places a patient in a high-risk group for cerebrovascular diseases. From a mechanical standpoint, the elastic limit of blood vessels is dictated by the underlying network of collagen fibers that are ultimately responsible for the representative elastic properties.[2] Thus, the overall elasticity of a vessel is substantially influenced by the presence and metabolism of collagen for maintaining the structural integrity of the vessel wall.

3.3.3 Physiological Properties of Vessels

Under normal physiological circumstances, circulatory function is effectively and efficiently regulated by internal feedback mechanisms of the heart and blood vessels. However, when subject to external conditions or stimuli such as temperature, emotional behavior, or drugs, the human body reacts by increasing or decreasing the heartbeat corresponding to an increase or decrease in blood flow. This internal mechanism, known as autoregulation, ensures the required volume of oxygenated blood and rate of perfusion to adjacent tissue and organs.

In humans and other mammals, multiple cardiovascular regulatory mechanisms have evolved to increase blood supply to active tissues and increase or decrease heat loss from the body by redistributing the blood. Circulatory adjustments are effected by local and systemic mechanisms that change the caliber of the arterioles and other resistance vessels and therefore change the hydrostatic pressure in the capillaries. The systemic mechanisms are both chemical and neural. A discussion of these is beyond the scope of this book, except to say that brain vessels receive vasoconstrictor and vasodilatory nerve fibers from neural sources inside and outside the brain. The systemic mechanisms act in synergy with local mechanisms and adjust vascular responses throughout the body. The local mechanisms maintain blood flow despite fluctuations in perfusion pressure by dilating the arterioles in active tissues.

Autoregulation is the term denoting the capacity of tissues to regulate their own blood flow. Most vascular beds have an intrinsic capacity to compensate for moderate changes in perfusion pressure by changes in vascular resistance, so that blood flow remains relatively constant. Auto-regulation is well developed in normal brain arteries. It is thought to occur in part due to the intrinsic contractile response of muscle to stretch (the myogenic theory of autoregulation). As the blood pressure rises, the vessel is distended, and in response its component muscle fibers contract. The higher the pressure, the greater is the degree of contraction. This is necessary to maintain the same wall tension (by applying the law of Laplace). Vaso-dilator substances also accumulate in active tissues, and these metabolites also affect autoregulation (metabolic theory of autoregulation). When blood flow decreases, metabolites accumulate and the vessels dilate; when flow increases, they are washed away and the vessels constrict.

Autoregulation is a physiological mechanism that is subject to internal thresholds beyond which it is rendered ineffective. The influence of auto-regulation and its inherent properties on the prevalence and clinical course of cerebrovascular disease has yet to be fully elucidated in the medical liter-ature. An example of the issues involved in the proper characterization of autoregulation and its potential role in cerebrovascular diseases will be described in detail in the chapter devoted to the physics of arteriovenous malformations (Chap. 7).

3.4   Circulatory Network: Specific Considerations

The primary objective of the circulatory system is to maintain adequate perfusion of the various tissues and organs of the human body. The heart accomplishes such a task under an oscillating systemic pressure gradient, sufficient to drive the blood flow through the circulation. In this section, we address the physiological component and related aspects of the circulatory system.

3.4.1   Blood Pressure

In addition to blood flow, the most important parameter in the circulation is the blood pressure. The blood pressure must be sufficient to drive the blood from the heart and into the blood vessels but must also be low enough to create a pressure gradient and allow efficient draining of the blood back to the heart for additional cycles. As one can imagine, the blood pressure varies significantly at various points in the circulation, allowing the different blood vessels to accommodate their assigned role and function. The left ventricle ejects blood from the heart into the arterial system under a pressure of approximately 120 mm Hg. This corresponds to the systolic phase of the cardiac cycle and is commonly referred to as systolic blood pressure. As blood penetrates further into the circulatory system, the pressures start to decline steadily. From the aorta, blood flows through the larger arteries at 110 mm Hg, through the medium arteries at 75 mm Hg, and through the smaller arteries or arterioles at 40 mm Hg until it reaches the capillary bed. Blood enters the capillary bed under a pressure of 30 mm Hg and exits under a pressure of 16 mm Hg. The low pressure is sufficient to generate movement while slow enough for the blood to perform its physiological function of sustaining cellular metabolism. Blood drains from the capillary bed into the smallest veins or venules at 16 mm Hg, continuing into the medium-sized veins under a pressure of 12 mm Hg and into the large veins at 4 mm Hg before reentering the heart for another circulatory cycle. The pressures within the circulatory system range from 120 to 4 mm Hg and maintain this range on a continual basis.

Working in a periodic, pulsatile motion, the function of the heart can be represented as a wave. At the peak of the wave, the heart ejects blood into the circulation under systolic blood pressure. However, there exists a phase in the cardiac cycle where the heart is relaxing while the atria fills with pooling blood. The heart enters the relaxation period under much less pressure than the systolic blood pressure. The relaxation phase of the cardiac cycle is referred to as diastole and the pressure maintained by the heart during this phase is the diastolic blood pressure. The diastolic blood pressure for a normal person is 80 mm Hg. During a physical exam, the systemic blood pressure is typically presented as

$$(\text{Systemic blood pressure}) = \frac{\text{Systolic blood pressure}}{\text{Diastolic blood pressure}} = \frac{120 \text{ mm Hg}}{80 \text{ mm Hg}}$$

In most cases, it is more convenient to condense these two blood pressure readings into a single one that also represents the overall status of the blood pressure in a patient. This is accomplished through a mean blood pressure, BP_{mean}, defined by

$$BP_{mean} = (\text{Systolic BP}) + \frac{2 \times (\text{Diastolic BP})}{3}.$$

3.4.2   Arterial Pulse

The blood pumped into the aorta during systole not only moves the blood in the vessels forward but also sets up a pressure wave that travels along the arteries. The pressure wave expands the arterial walls as it travels, and the expansion is palpable as the pulse. The pulse wave is independent of and much faster than the velocity of blood flow. Its rate of travel is about 4 m/s in the aorta, 8 m/s in large arteries, and 16 m/s in small arteries of young adults.

3.4.3   Resistance and Capacitance Vessels

The veins are an important blood reservoir. Normally they are partially collapsed and elliptical in cross section. A large volume of blood can be added to the venous system before the veins become distended to the point where further increments in volume produce a large rise in venous pressure. The veins are therefore called the capacitance vessels. Small arteries and arterioles are called resistance vessels because they are the main site of peripheral resistance.

At rest, 50% of the circulating blood volume is in the systemic veins. 12% percent is in the heart chambers, and 18% is in the low-pressure pulmonary circulation. Only 2% is in the aorta, 8% in the arteries, 1% in the arterioles, and 5% in the capillaries.

3.4.4   Arterial Blood Flow and Pressure

The velocities and pressures of blood in the various parts of the systemic circulation are summarized in Fig. 3.4. The mean velocity of blood in the proximal portion of the aorta is 40 cm/s. This, however, is phasic (pulsatile) and ranges from 120 cm/s during systole to a negative value at the time of the transient backflow before the aortic valves close in diastole. In the distal aorta and in other large arteries, the velocity of blood is greater in systole than it is in diastole. However, the vessels are elastic, and forward flow is continuous because of the recoil during diastole of the vessel walls that have been stretched during systole. This recoil effect is called the *Windkessel* effect (German for "elastic reservoir").

The pressure in the aorta and other large arteries rises normally to a peak value (systolic pressure) of about 120 mm Hg during each cardiac cycle and falls to a minimum value (diastolic pressure) of about 70 mm Hg. The arterial blood pressure is expressed as systolic pressure over diastolic pressure, e.g., 120/70 mm Hg. 1 mm Hg is equivalent to 0.133 kPa. The *pulse pressure* is the difference between systolic and diastolic pressures and is therefore normally about 50 mm Hg. The *mean pressure* is the average pressure throughout the cardiac cycle. As systole is shorter than diastole, the mean pressure is slightly less than the value halfway between systolic and diastolic pressure. It can be determined only by integrating the area under a blood pressure–time curve; however, as an approximation it is the diastolic pressure plus one-third of the pulse pressure. The mean pressure falls considerably at the end of the arterioles to 30–38 mm Hg. Gravity also has an effect on blood pressure; the pressure in any vessel below heart level is increased and that in any vessel above heart level is decreased by the effect of gravity. Arterial pressure is the product of the cardiac output and the peripheral resistance, and therefore it is affected by either or both of these factors. In general, increases in cardiac output increase the systolic pressure, and increases in peripheral resistance increase the diastolic pressure.

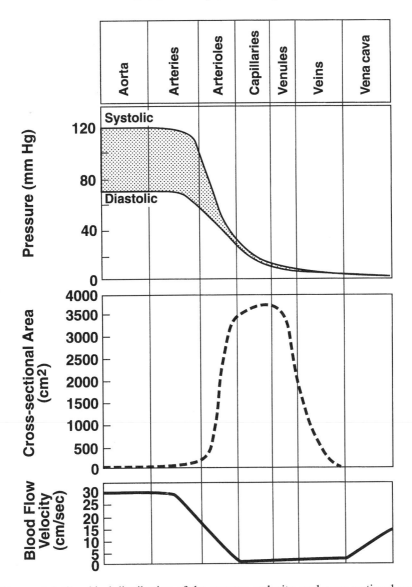

FIGURE 3.4. Graphical distribution of the pressure, velocity, and cross-sectional area of the primary components of the human circulatory system.

Capillary pressures vary considerably, but typical values in human nail bed capillaries are 32 mm Hg at the arteriolar end and 15 mm Hg at the venous end. The capillary hydrostatic pressure is opposed by an osmotic pressure exerted by plasma proteins since they cannot readily permeate the vessel wall in sufficient amounts. The osmotic pressure is maintained at an average value of 15 mm Hg and results in an overall effective driving force of 10 mm Hg toward the arteriole end of the capillary vessel bed, driving water and electrolytes out into tissue fluids, and an effective pulling force of 15 mm Hg, drawing water and electrolytes from tissue fluids into the blood stream.[3] The pulse pressure is about 5 mm Hg at the arteriolar end and 0 at the venous end. The capillaries are short, but blood moves slowly (about 0.07 cm/s) because the total cross-sectional area of the capillary bed is large. Transit time from the arteriolar to the venular end of an average sized capillary is 1–2 s. The pressure in the venules is 12–18 mm Hg; it falls steadily in the larger veins to about 5.5 mm Hg in the great veins outside the thorax. At the entrance to the heart the *central venous pressure* averages 4.6 mm Hg but fluctuates with respiration and heart action. The effect of gravity increases the pressure by 0.77 mm Hg for each centimeter below the right atrium and decreases it by a similar amount above the right atrium. In the great veins, the velocity of blood is about one-fourth that in the aorta, with an average of about 10 cm/s.

The discussion on arterial blood flow and pressure in this section was intended to introduce the reader to the magnitude and variability of these values throughout the circulatory system. The role of these two parameters in the onset and development of cerebrovascular diseases cannot be overstated and will be the sole topic presented in the chapter on the physics of blood flow or hemodynamics (Chap. 4).

3.4.5   Methods for Measuring Blood Pressure

As was discussed previously, the pressures within the arteries at any given moment vary considerably between the maximum at systolic pressure (≈ 120 mm Hg) to the minimum at diastolic pressure (≈ 80 mm Hg). The arterial blood pressure in humans is measured routinely and noninvasively by the auscultatory method, using an inflatable cuff attached to a mercury manometer (the sphygmomanometer). The cuff is wrapped around the arm and inflated to above the expected systolic pressure. At this point, the artery is totally constricted due to the pressure exerted by the inflatable cuff being greater than the pressure within the artery. Also, the manometer responds with a rise in the column of mercury. A stethoscope is placed over the brachial artery and, as the cuff is deflated, the sounds of the blood are heard spurting through the brachial artery as it opens, indicating the systolic pressure. As the cuff pressure is lowered further, the sounds become louder, then muffled (at the diastolic pressure), and finally they disappear. These sounds are produced by turbulent flow in the brachial artery. When the artery is narrowed, the velocity of flow through the constriction exceeds the *critical velocity* and turbulent flow results.

The range of pressures recorded between cardiac systole and diastole, however, do not accurately reflect the precise pressure measurement within a given vessel. Particularly in applications of cerebrovascular disease, diagnosis and treatment typically involve the use of catheters and injection of either contrast medium or embolic agent or surgical clipping, which, in turn, introduces perturbations and instabilities of the pressure distribution within the local vessels. These pressure measurements can be accommodated through a more invasive approach than that of the sphygmomanometer. The arterial pressure may be measured directly if a cannula or catheter inserted into an artery is connected to a mercury manometer or a suitably calibrated strain gauge and recording device. In arteries of the brain, this may be done directly during surgery after exposing the artery in question, or it may be performed in a minimally invasive manner by navigating a thin caliber hollow angiographic microcatheter (which in effect acts as a very thin long cannula) through the arterial system of the body (usually through a small hole in an artery in the groin) to the artery in question within the brain. In one study, the experimental evaluation of two microcatheter systems for intravascular pressure monitoring established the reliability of mean blood pressure measurements in both animal and human subjects.[4]

The primary objective of intravascular pressure measurements using a microcatheter is to harness and transfer the energy created by the hemodynamic forces acting on the catheter to a recording or detection device that would allow direct measurement of the pressure. As the blood interacts with and enters the microcatheter, it induces a vibrational displacement of an elastic membrane attached to the opposing end of the catheter. Thus, in this method of measurement, the pressure can be related and subsequently quantitated by relating it to the frequency of the vibrating elastic membrane. Such a relation can be derived by analyzing the forces acting on the catheter–manometer system, which are dependent on the frequency of the vibrating elastic membrane and solving the equation for the frequency in terms of the system variables. The force due to the pressure required to propel the blood up and down through the catheter, F_p, is dependent on the sum of three physical forces: F_a, force due to the acceleration of the mass of the manometer system; F_v, viscous force exerted by the fluid in motion through the catheter; and F_e, elastic force required to distend the membrane. These forces can be expressed in a mathematical relation based on Newton's laws of motion:[5,6]

$$P = M \frac{d^2 X}{dt^2} + C \frac{dX}{dt} + kX,$$

where P is the pulsatile pressure driving blood through the manometer, M is the effective mass of the elastic membrane ($= \rho L$, ρ is the density of blood, L is the length of the catheter), C is an effective viscoelastic damping coefficient ($= 8\pi L/R^2$, R is the radius of the catheter), K represents the effective spring constant of the elastic membrane ($= E\pi R^2$, E is the elastic modulus of the membrane), and X is the vibrational displacement. Solution of this equation yields an expression for the vibrational displacement in terms of the system parameters. This, in turn, provides a quantitative value for the pressure causing the vibrational displacement of the membrane. This

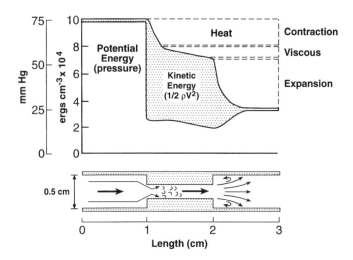

FIGURE 3.5. Illustrative example of Bernoulli s principle, depicting the distribution of kinetic energy and potential energy in a vessel containing a constriction or stenosis. (Source: Sumner DS. Essential hemodynamic principles. In: Rutherford R.B. (ed). *Vascular Surgery*, fourth edition. Philadelphia: W.B.Saunders Company; 1995: p. 24. Reprinted with permission.)

equation represents a common problem in physics and engineering of a damped vibrating system and is described in detail in other textbooks.[7,8]

3.5 Bioenergetics of the Circulation

As the heart continually exerts a pulsatile force in propelling the blood through the blood vessels of the circulatory system, it is also performing energy, work, and power. A physical system representing the motion of a fluid can be described by Bernoulli's principle, which, in effect, is an extension of the conservation of energy. Bernoulli's principle, applied typically to fluid flow in vessels exhibiting sudden contractions or expansions such as those presented by an arterial stenosis or the carotid sinus (Chap. 5), is illustrated in Fig. 3.5. A fluid, in nature, is subject typically to two types of energy: potential energy and kinetic energy. The *potential energy* of a fluid is the sum of the energy due to intravascular pressure P exerted on the fluid starting it in motion and the gravitational potential energy (GPE). Because the intravascular pressure induces fluid motion, it possesses the ability to perform useful work[9] and is thus the primary source of potential energy. GPE applies to fluids subjected to various heights above baseline level. The fluid then can be described as having gravitational potential energy GPE as

$$GPE = \rho gh,$$

where g is the acceleration due to gravity, and h is the distance or height above baseline levels.

A fluid in motion, as is the case with any object in motion, exhibits *kinetic energy*. For a solid object, the kinetic energy, (KE), is

$$\text{KE} = \tfrac{1}{2}mv^2.$$

For a fluid, the kinetic energy assumes the same form with its mass expressed in terms of its density or

$$\text{KE} = \tfrac{1}{2}\rho Vv^2,$$

where ρ is the fluid density and V is fluid volume. The total energy E exerted by a fluid is the sum of its kinetic energy KE, potential energy PE, and gravitational potential energy GPE and is constant, according to

$$E_{\text{tot}} = \text{KE} + \text{PE} + \text{GPE} = \text{const},$$

$$E_{\text{tot}} = P + \tfrac{1}{2}\rho Vv^2 + \rho gh = \text{const}.$$

If, for example, flow in an artery is interrupted or slowed by an occlusion, all the kinetic energy of flow is converted into pressure energy. Thus, when an artery is tied off beyond the point at which the cannula is positioned, an *end pressure* is recorded. If pressure is measured at a point in the artery without interruption of flow (i.e., no induced resistance) then the recorded *side pressure* is less than the end pressure by the kinetic energy of flow. This is because in a tube or a blood vessel the total energy (the sum of the kinetic energy of flow and the pressure energy) is constant (Bernoulli's principle). When pressure drops in any segment of the arterial system it is due both to resistance and to conversion of potential into kinetic energy. The pressure drop due to energy lost in overcoming resistance is irreversible, since the energy is dissipated as heat; but the pressure drop due to change or transformation of potential to kinetic energy as a vessel narrows is reversed when the vessel widens out again. Bernoulli's principle is an extremely important physical concept and will be discussed further in Chap. 4 with applications to stroke presented in Chap. 5 and intracranial aneurysms presented in Chap. 6.

3.6 Analog Models of the Circulation

Attempts to investigate various aspects of the circulation system in any capacity are limited primarily by the complexity of the structure and arrangement of the associated blood vessels and related physiological mechanisms. The difficulty in such research endeavors can be alleviated to some degree through the formulation and development of theoretical models. This permits not only accurate renderings of approximations of such complex physiological processes but also allows the researcher to simulate the effects of abnormal conditions by inducing appropriate perturbations in such models without laborious experimentation. One approach to simulating the circulation is through the use of analog models or simplified models of circulatory blood flow based on similarities of fluid flow found in the physical sciences.

3.6.1   Windkessel Model

The *Windkessel* model, introduced by German physiologist Otto Frank in 1899, is one of the first attempts at modeling the circulation. The model consists primarily of an elastic reservoir to simulate the heart or cardiac chamber connected by two elastic vessels: one representing arterial blood flow into the reservoir and the other representing venous blood flow out of the reservoir. The pressure and volume inside the reservoir were assumed to be proportional to each other according to

$$(\text{Volume}) = (\text{Compliance}) \times (\text{Pressure}),$$

where the proportionality constant is the compliance of the elastic reservoir or its change in volume with respect to pressure. The objective of this model is to relate the change in volume of the elastic reservoir, dV/dt, to its corresponding blood flow Q, or

$$\frac{dV}{dt} = Q. \tag{3.1}$$

In physiological approximations, there are two types of flow that govern circulation from the heart. The first flow, Q_p, is due to the blood currently within the reservoir that flows linearly with respect to the pressure P and resistance R, according to Poiseuille's law (Chap. 4):

$$Q_p = \frac{P}{R}. \tag{3.2}$$

The second flow, Q_s, accounts for the additional surge of blood from the heart at cardiac systole and is described mathematically as

$$Q_s = C\frac{dP}{dt}, \tag{3.3}$$

where C is the capacitance of the elastic reservoir. The sum of Q_p and Q_s represents the total flow through the reservoir and is substituted into Eq. (3.1) to yield the differential equation describing circulation in the *Windkessel* model:

$$Q = C\frac{dP}{dt} + \frac{P}{R}. \tag{3.4}$$

This is a second-order differential equation in terms of the pressure within the reservoir, which suggests a possible solution in exponential form,[10]

$$P = P_0 e^{-t/RC},$$

and, upon substitution into Eq. (3.4), yields the exact solution for the pressure:

$$P = e^{-t/RC} \left(P_0 + \frac{1}{C} \int_0^t e^{-\tau/RC} Q(\tau) d\tau \right),$$

where P_0 is the original pressure. If the function of the flow wave form is known, it can be substituted into the above equation for determination of the pressure at any given instance during the cardiac cycle.

3.6.2   Electrical Analogs

As a physical concept, fluid flow can be difficult to comprehend. In the simplest case, fluid flow through a rigid tube is driven by a difference of pressures between the two ends of the tube or a pressure gradient (ΔP). The fluid is subject to a resistance force (R) dictated by the tube radius, tube length, and viscosity of the fluid. In terms of the pressure gradient and resistance, the volumetric rate of fluid flow Q is given by

$$Q = \frac{\Delta P}{R}.$$

This is known as Poiseuille's law and will be explained in detail in Chap. 4. For an isolated rigid tube, Poiseuille's law explains fluid flow in a logical and straightforward manner.

However, in its crudest form of approximation, vascular structure consists of a multiple arrangement of interconnected tubes with various lengths and radii. The determination of flow in each of these vessels becomes a more complex problem than the case of flow through the isolated vessel. One simply cannot calculate flow in each vessel of the arrangement by isolating them and applying Poiseuille's law without accounting for the contribution of pressure and resistance from interconnected vessels. This obstacle has thus prompted the search for physical analogs or complementary techniques that can be applied to the problem of fluid flow in an accurate and efficient manner.

One such analog follows from the principles of electricity and electrodynamics. In comparison to an example described previously, fluid flow through a rigid tube is analogous to current flow through a wire and thus an electrical analogy can be introduced. Using an electrical analog, current is analogous to fluid flow rate, voltage is analogous to the pressure gradient, resistance remains the same concept in both cases, conductance is analogous to vascular compliance, and inductance is analogous to inertia. Table 3.2 summarizes the fluid flow parameters and corresponding electrical entities.

TABLE 3.2. Electrical analogs of hemodynamic parameters.

Hemodynamic Parameter	Electrical analog
Volumetric blood flow rate Q (cm^3/s)	Current I (A)
Pressure gradient ΔP (dyn/cm^2)	Voltage V (volt)
Vascular resistance R (dyn s/cm^5)	Resistance R (Ω)
Inertance L (g/cm^4)	Inductance L (henry)
Compliance C (cm^5/dyn s)	Capacitance C (farad)

As the resistance increases in a wire, the current will be reduced just as a corresponding increase in a tube will decrease the fluid flow rate. Thus, employing common electrical principles, it becomes a simple task to investigate flow through an isolated tube.

More importantly, electrical principles are often applied to the calculation of current, not primarily through single wires, but through networks of wires connected as circuits. The multiple arrangement of tubes representing the vascular system, in effect, resembles a complex circuit. Consider the simplified arrangement of connected tubes, resembling an electrical network circuit, presented in Fig. 3.6. The circuit consists of an arrangement of connected wires (tubes) of various resistances R, powered by a voltage source V (pressure gradient ΔP). The primary goal in solving this network is to determine the total and the individual current (flow) through each wire (tube) in the circuit. However, each of the electrical variables behaves differently, depending on whether the wires in the circuit are connected in parallel or in series. The effective resistance of three wires connected in series, $R_{\text{eff},s}$, is given by

$$R_{\text{eff},s} = R_1 + R_2 + R_3,$$

while the effective resistance of three wires connected in parallel, $R_{\text{eff},p}$, is

$$\frac{1}{R_{\text{eff},p}} = \frac{1}{R_1} + \frac{1}{R_2} + \frac{1}{R_3},$$

which can be rearranged to yield

$$R_{\text{eff},p} = \frac{R_1 R_2 R_3}{R_1 + R_2 + R_3}.$$

One may extend this concept in creating much more comprehensive models of the circulation by introducing additional electrical components that exhibit other common hemodynamic properties. For example, from the electrical components and their corresponding hemodynamic counterparts introduced earlier, vascular compliance can be represented by including a capacitor into the electrical circuit, and inertial effects can be represented by including an inductor into the electrical circuit. In terms of hemodynamically relevant factors, the vascular resistance R_v, capacitance C, and in-

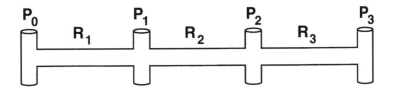

Pressure Gradient: $\Delta P_T = (P_0 - P_3) = (P_0 - P_1) + (P_1 - P_2) + (P_2 - P_3)$

Resistance: $R_T = \dfrac{\Delta P_T}{Q_T} = \dfrac{(P_0 - P_3)}{Q_T} = R_1 + R_2 + R_3$

Flow: $Q_T = Q_1 + Q_2 + Q_3$

Resistance: $\dfrac{1}{R_T} = \dfrac{Q_T}{\Delta P_T} = \dfrac{(Q_i - Q_f)}{(P_i - P_f)} = \dfrac{1}{R_1} + \dfrac{1}{R_2} + \dfrac{1}{R_3}$

$$= \dfrac{R_1\, R_2\, R_3}{R_2 R_3 + R_1 R_3 + R_1 R_2}$$

FIGURE 3.6. Elementary physiological principles of fluid flow through tubes or blood vessels connected in (A) series and (B) parallel.

ductance L are defined by[10]

$$R_v = \frac{8\pi\mu L}{S^2},$$

$$C = \frac{3S(1+h)^2}{E(1+2h)},$$

$$L = \frac{\rho L}{S},$$

where S is the cross-sectional area of the vessel, L is the vessel length, μ is the fluid viscosity, ρ is the fluid density, E is Young's modulus of the vessel wall, and h is the radial to wall thickness ratio.

Solution of the electrical circuits containing these aforementioned components yields the following equations, known as the telegraph equations:

Electrical	Hemodynamic
$\dfrac{\partial V}{\partial z} = L\dfrac{\partial i}{\partial t} + iR$	$\dfrac{\partial P}{\partial z} = L\dfrac{\partial Q}{\partial t} + QR_v$
$\dfrac{\partial i}{\partial z} = C\dfrac{\partial V}{\partial t} + \dfrac{V}{R}$	$\dfrac{\partial Q}{\partial z} = C\dfrac{\partial P}{\partial t} + \dfrac{P}{R_v}$

The telegraph equations have been used to successfully investigate the hemodynamic effects of the circulation[11,12] and in the development of biomathematical models to describe qualitatively the influence of pulsatile pressure forces on the dynamics and growth progression of intracranial aneurysms[13] (Chap. 6). Electrical circuits have also been applied to the investigation of hemodynamics of arteriovenous malformations (Chap. 7).

3.7   Circulatory Network: Cerebral Circulation

Up to this point, our discussion of the circulation system has centered on general considerations of the anatomy of the vasculature and the physiological function of the heart. However, we now focus our attention on specific considerations of the vascular arrangement and circulatory physiology within the brain and their relevance to the onset and development of cerebrovascular diseases.

3.7.1   Vascular Supply of the Brain

The two brain hemispheres are supplied with blood from four arteries, two internal carotid arteries and two vertebral arteries. Their arterial branches are directed to the gray matter of the brain, which needs more blood than the white matter. An artery that enters the surface of the brain is always an end artery, and therefore infarction (death) of neuronal tissue may occur following an obstruction of the artery.

The internal carotid and vertebral systems anastomose or join with each other at the base of the brain, forming the circle of Willis (which is actually polygonal in shape). Posteriorly, the two vertebral arteries fuse to form a single basilar artery. The cerebrovasculature, as well as components and branches of the circle of Willis, is shown schematically in Fig. 3.7. The circle of Willis is so constantly present that it suggests there is some functional reason for its existence. A possible reason is that it equalizes the pressure and the volume of blood flow between the two sides of the brain. Normally little mixing of blood from opposing streams occur in the communicating arteries of the circle of Willis because the pressures of the two streams are equal. Should either vertebral or carotid system be occluded, redistribution (clinically, termed *cross-filling*) of blood occurs through the circle of Willis.

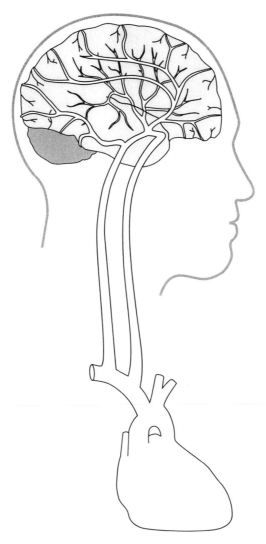

FIGURE 3.7. Schematic diagram of the arrangement of cerebral blood vessels, particularly those comprising the circle of Willis, located at the base of the brain.

The arterial supply of the cerebral cortex is by the three cerebral arteries (anterior, middle, and posterior) branching off the circle of Willis. The former two arise from the internal carotid system and the posterior cerebral artery is the terminal branch of the basilar artery. The branches of these three arteries anastomose across the borders of their respective territories and on the surface of the brain, but very sparsely. Their *perforating arteries* are end arteries, as explained above. The middle cerebral artery is the largest

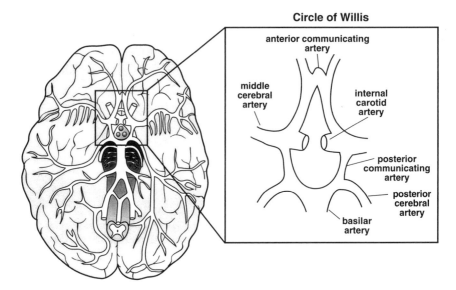

FIGURE 3.7 (*continued*)

and most direct branch of the internal carotid artery. It feeds most of the lateral aspect of the cerebral cortex and gives off central branches to supply the internal capsule that contains the motor and sensory neural fibers supplying most of the opposite side of the body. Venous drainage from the brain occurs by way of the deep veins and dural sinuses, emptying mainly into the internal jugular veins.

3.7.2 Cerebral Blood Flow

Regulation of the cerebral circulation generally maintains a constant total cerebral blood flow (CBF, the amount in milliliters perfusing 100 g of brain tissue per minute) under varying conditions. Regional cerebral blood flow (rCBF) is determined by the ratio of regional cerebral perfusion pressure (rCPP) and regional cerebrovascular resistance (rCVR) by[14]

$$rCBF = \frac{rCPP}{rCVR}$$

and can be obtained through a variety of imaging techniques (Fig. 3.8). Factors affecting the total CBF include the arterial and venous blood pressures, the viscosity of blood, the intracranial pressure, and the degree of constriction or dilatation of the cerebral arterioles.[15] The latter is controlled by neuronal stimuli, cerebral metabolism, and autoregulation.

As a consequence of autoregulation, CBF remains constant over a range of blood pressures, which in the normotensive person extends from a *mean*

CT   MRI   CBF   CMRGl
                      ml/100g/min   μmol/100g/min

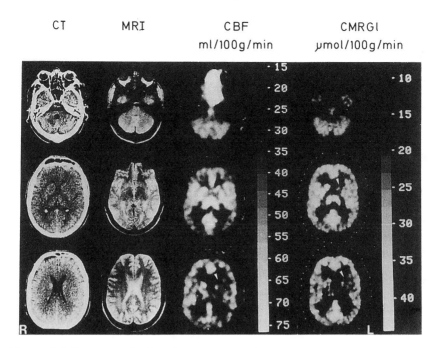

FIGURE 3.8. Representative imaging modalities used to acquire cerebral blood flow measurements. (Source: Toole JF. *Cerebrovascular disorders, Fourth edition.* New York: Raven Press; 1990: p. 198. Reprinted with permission.)

pressure of 130 mm Hg (above which blood flow "breaks" through autoregulation) down to a *mean* pressure of 60 mm Hg. These are called the upper and lower limits of cerebral autoregulation, as illustrated in Fig. 3.9.[16,17] In this normal range, the CBF is 50–55 mL per 100 g/min. Once the blood pressure falls below the lower limit of autoregulation, the CBF falls passively in parallel with pressure. The blood pressure at which ischemia occurs is considerably below the lower limit of autoregulation. Values of CBF < 30 mL per 100 g/min (at about 30 mm Hg arterial blood pressure) have been shown to correlate with an elevated risk of ischemia or stroke. Perfusion pressure correlates with mean pressure (not systolic) because it more closely represents the active perfusion driving force.

3.7.3   Experimental Models of the Cerebral Circulation

In its most simplified form, blood flow through a blood vessel can be recreated and thoroughly investigated by considering the flow of a fluid through a rigid tube. Due to the easy availability of all necessary equipment (assembly of rigid tubes, viscous fluid, and a fluid pump), it is possible to represent segments and resultant hemodynamics of the human circulation.

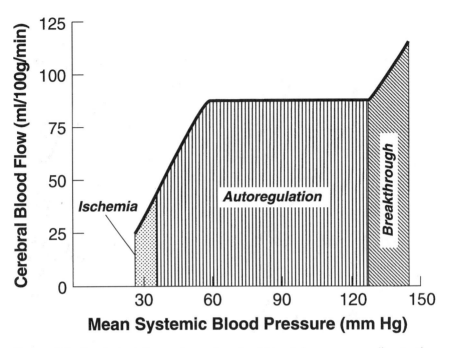

FIGURE 3.9. Graph depicting regions of cerebral blood flow corresponding to the physiological phenomena of ischemia, autoregulation, and breakthrough.

Applications of engineering analysis have been applied experimentally to the study of hemodynamics within the circle of Willis.[18-23] The circle of Willis, as will be shown in later chapters of this book, is the region of the cerebral circulation where the origin of cerebrovascular aneurysms are prevalent.

3.7.4   Measurement of Cerebral Blood Flow

Quantitative knowledge of cerebral blood flow allows one to assess accurately the physiological state of cerebral physiology and function. Minimally invasive techniques for the measurement of cerebral blood flow placing the patient at minimum risk involve imaging procedures that provide physiological information in the acquired images. The imaging techniques that are currently employed clinically to obtain cerebral blood flow measurements are transmission computed tomography (TCT, x-ray CT) and emission computed tomography [ECT, single-photon emission computed tomography (SPECT) and positron emission tomography (PET)]. The energy sources utilized in these imaging modalities to produce three-dimensional images of the brain circulation are x rays or radioisotopes.

The premise behind the measurement of regional cerebral blood flow is to image the volumetric distribution of a tracer or a chemical substance that is physiologically inert, i.e., does not interact with normal physiological processes within the brain, but also can interact with the energy source (x rays or radioactive emissions) so as to produce visual maps of the source distribution. A radioisotope is a unique type of radioactive source or tracer that is a single-photon emitter or emits photons as a primary emission upon decay. The tracer is administered intra-arterially or intravenously, usually through injection, at a vascular point proximal to the brain. With regard to tracer administration, the only exception involves the use of ^{133}Xe as a tracer. ^{133}Xe is a chemically inert gas that can be inhaled in addition to injected prior to imaging. As the circulation carries the tracer through the blood vessels of the brain, images are acquired as the tracer distributes itself within the brain and then passes through the venous circulation. The general relation used in the measurement of cerebral blood flow was initially derived as[24,25]

$$T_t = \frac{V}{F},$$

where T_t is the mean transit time, V is the volume of tracer distribution, and F is the blood flow through the brain. This equation is based on the principle of conservation of matter and states that the tracer particles introduced into a perfused organ must sooner or later be washed out.[26] Specific examples of clinical techniques for the measurement of cerebral blood flow will now be introduced and briefly described.

3.7.4.1   Nitrous Oxide Inhalation

In this technique, first introduced by Kety and Schmidt,[27] application of the Fick principle permits measurement of cerebral blood flow by determining the amount of a given substance (e.g., inhaled nitrous oxide) removed from the blood stream by the organ per unit time and dividing that value by the difference between the concentration of the substance in arterial blood and that in venous blood of the brain. Levels of nitrous oxide are measured and quantitated by hand-held detectors at surgery. The value obtained by the Kety–Schmidt method is an average value for flow during a 10-min equilibration period. Furthermore, the values obtained for total CBF give no information about the relative blood flow to various parts of the brain.

3.7.4.2   Computed Tomography

In computed tomography (CT), x rays are used to obtain two-dimensional (2D) projections of the cross-sectional area of an object or anatomical region of interest. In CT, the x-ray tube is affixed to a circular gantry that rotates about the region of interest for a given patient collecting 2D projections at user-defined angular increments. These 2D projections are then reconstructed mathematically to produce a 3D image of the object of interest. The application of CT imaging to cerebral blood flow measurements involves the interaction of the x-ray beam with tracers, increasing the attenuation of x rays through regions of the brain perfused with the tracer. Specifically, the various techniques for cerebral blood flow measurement involve the administration of the following tracers: nitrous oxide and ^{133}Xe.

3.7.4.2.1   Intra-Arterial Injection of ^{133}Xe

Using the intra-arterial ^{133}Xe method, the blood flow of different regions of the brain can be measured by injecting the radioactive gas ^{133}Xe dissolved in saline into a single artery such as the carotid artery.[28] The arrival and clearance of the tracer in various regions is monitored with an array of scintillation detectors placed over the head. Each detector is collimated to scan about 1 cm^2 of brain surface. Computer analysis and color display of the flow data is possible. This technique allows only the radioactivity from the cerebral surface to be detected. In a related technique, i.e., xenon computed tomography (Xe-CT), ^{133}Xe is administered and distributed through the cerebral circulation according to the principles described above. The primary exception is the means of detection with the tissue tracer concentration throughout the brain substance being monitored with computed tomography.

Determination of cerebral blood flow is performed in three distinct steps:[28]

1. *Quantitation of xenon concentration in the brain.* The concentration of xenon in milligrams per milliliter in a volume element of the brain ($C_{Xe,brain}$) is related to the ratio of the mass attenuation coefficient of water (μ_w) to that of xenon (μ_{Xe}) by

$$C_{Xe,brain} = HE \, \frac{\mu_w}{\mu_{Xe}},$$

where HE is a constant termed the Hounsfield enhancement.

2. *Quantitation of xenon concentration in blood.* The concentration of xenon in the blood is determined by the mathematical approximation of the time-dependent arterial xenon concentration $C_{Xe,art}$:

$$C_{Xe,art}(t) = C_{Xe,max}(1 - e^{-bt}),$$

where t is the time since the start of xenon inhalation and b is the rate constant for arterial uptake of xenon, calculated from the breathing rate and lung volume.

3. *Quantitation of cerebral blood flow.* The physical entities presented in the previous two categories are related through a modification of Fick's law of diffusion:

$$C_{Xe,brain}(t) = \lambda k \int_0^t C_{Xe,art}(u) e^{-k(t-u)} \, du,$$

where u is a temporary variable. The equation is solved simultaneously for

λ and k by applying the least-squares method to fit it to the volumetric element data in each image. Blood flow is represented by λk.

3.7.4.3   Single Photon Emission Computed Tomography

Single-photon emission computed tomography (SPECT) is a radiation detection imaging technique that is used to acquire qualitative and quantitative measurements of cerebral physiology. The presence of cerebrovascular disease introduces abnormalities in the values and patterns of regional and local cerebral blood flow which can readily be identified from SPECT. In a SPECT scan, a specific type of radioisotope is first administered into the bloodstream of the patient. As the radioactive source distributes itself and localizes within the body, it is continually emitting photons that are of sufficient energy to penetrate through surrounding tissues and escape the human body. The photons that escape the body are then registered and collected by a detector assembly. The detector assembly consists of a collimator that acts to restrict unwanted scattered photons from reaching the detector, thereby increasing image quality and sharpness. Once they pass the collimator, the photons impinge upon a γ-ray detector that is coupled to photomultiplier tubes. The two-dimensional position and intensity of the detected γ ray are registered to form a "snapshot" image of the object or organ. Snapshots are acquired at angular increments within a circular or elliptical orbit about the organ of interest. These snapshots are then reconstructed mathematically to provide the physician with a three-dimensional distribution of the activity distributed within the organ.

In applications of SPECT to cerebral blood flow measurements, regional CBF can be determined by monitoring the cerebral transit of inhaled ^{133}Xe or injected radionuclides ^{99m}Tc hexamethyl–propyleneamine oxime[29] (HMPAO) or ^{123}I N-isopropyl-p-iodoamphetamine (IMP) using tomographic imaging through the intact skull.[30] Arrays containing sodium iodide scintillation crystals are mounted in a configuration that rotates around the subject's head. Collimators define a number of tomographic sections. Voxel flow values are displayed in a 64×64 or 128×128 matrix using a 16-shade or -color scale that can be normalized to the highest flow value. Calculations of CBF in addition to other applications of brain SPECT techniques to CBF measurements are outlined further in comprehensive reviews.[31–33]

3.7.4.4   Positron Emission Tomography

Positron emission tomography (PET) is similar in principle to SPECT in that both tomographies are forms of nuclear medicine techniques involving the detection of radioactive emissions from a circulating radiotracer.[34–36] In addition to the advantage of acquiring physiological information concerning regional cerebral blood flow, PET radionuclides can be covalently attached to 2-deoxyglucose, a metabolic form of glucose in constant demand by brain tissue. Consequently, PET images reveal valuable quantitative information concerning biochemical processes. The PET radionuclide decays by emitting a positron that travels a short distance, approximately 1–2 mm, before a collision with an electron and subsequent annihilation. Annihilation results in two photons of equal energy, 511 keV, moving in a direction opposite from one another. With regard to the issue of collimation, PET utilizes electronic collimation, relying on the 180° ejection of annhilation photon pairs for accurate detection of these photons. The coincidence of two detectors situated opposite to each other defines a line along which the radioactive decay process occurred. The superposition of a very large number of these lines is evaluated by an image reconstruction program to generate a three-dimensional image of the organ or region being investigated.

One advantage of PET, in addition to SPECT, is its ability to measure cerebral metabolic utilization of oxygen and glucose.[37] Intra-arterial injection of the radiotracer 2-deoxyglucose labeled with ^{18}F yields the metabolism of brain tissue as evidenced by the concentration of emitted positrons, and therefore provides a reflection on the blood flow. Therefore, blood flow can be determined by tomographic methods through the intact skull, as outlined above. Further sequential inhalation of ^{15}O-labeled O_2, CO_2, and CO also yields non-invasive methods of measuring cerebral blood volume (CBV), CBV to CBF ratio (mean vascular transit time), net extraction of oxygen, and cerebral oxygen and glucose metabolism.

3.7.4.5 Magnetic Resonance Imaging

Magnetic resonance imaging (MRI) is an imaging modality based on the nuclear phenomena of magnetic resonance. The structure of a typical atomic nucleus consists of neutrally charged neutrons and positively charged protons while negatively charged electrons are in continual orbital motion about the nucleus. Atoms possessing a nucleus with an even number of protons are in nuclear equilibrium, while those atoms possessing a nucleus with an odd number precess or spin about an axis in a spatially random orientation. Examples of such nuclei are ^{1}H, ^{13}C, ^{19}F, and ^{31}P. More importantly, the charge distribution between the intimately spaced entities (protons and neutrons) creates a magnetic moment and thus allows one to visualize the nuclear structure as a region of magnets randomly oriented in space and in constant precession.

When subjected to a uniform magnetic field, the magnets conform to the field lines established by the magnetic field, resulting in the spatial alignment of the magnets. If a pulsating beam of radio-frequency (RF) energy is focused onto the nucleus, energy is transferred to the spinning protons, and thus these spatially oriented magnets or magnetic moments flip from their lower-energy state to their higher-energy state and have now become aligned along a central axis. At the completion of an RF pulse, the magnetic moments return to their lower-energy state. The frequency required to induce the nuclear transitions of the magnetic moments between the lower- and higher-energy states is detected by an RF receiver.[38] The signals recorded from the RF receiver are then processed mathematically using Fourier analysis techniques to yield a visual image of the regions of tissue with characteristic frequencies.

Cerebral blood flow measurement techniques based on magnetic resonance imaging involve the perfusion of tissue and may be measured by detecting movements of protons.[39-43] This is done by designing pulse sequences that are sensitized to very small amounts of proton motion. Alternatively, rapid acquisition of images as an injected MR contrast agent passes through the brain provides information related to physiological perfusion.

It should be noted that there exist methods of cerebral blood flow measurement in addition to those mentioned previously such as ultrafast computed tomography[44] and color duplex sonography.[45] Space limitations prohibit an in-depth discussion of these experimental techniques and the reader is referred to the accompanying references for further reading.

3.8   Summary

In summary, this chapter addressed briefly the anatomical structure, arrangement, and physiological function of the circulatory system, in general, and the cerebral circulation, in particular. The most important biophysical characteristics of the circulation are blood pressure and blood flow. It is their behavior and interactions with the various blood vessels that illustrate possible mechanisms for the onset of cerebrovascular diseases. Also discussed were experimental and clinical techniques for obtaining measurements of blood pressure and blood flow. Understanding the normal structure and physiology sets the scene for appreciating the physical effects implicated in the development, detection, and treatment of cerebrovascular diseases.

3.9   References

1. A. Van Heuvelen, "Physics of the circulatory system," Phys. Teacher **27**, 590–596 (1989).
2. J.M. Gosline, "The physical properties of elastic tissue," Int. Rev. Connect. Tiss. Res. **7**, 211–249 (1976).
3. A.B. McNaught, *Illustrated Physiology* 3rd ed. (Churchill Livingstone, Edinburgh, 1976), p. 98.
4. G. Duckwiler, J. Dion, F. Viñuela, B. Jabour, N. Martin, and J. Bentson, "Intravascular microcatheter pressure monitoring: Experimental results and early clinical evaluation," Amer. J. Neuroradiol. **11**, 169–175 (1990).
5. D.J. Patel and R.N. Vaishnav. *Basic Hemodynamics and its Role in Disease Processes* (University Park, Baltimore, 1980), p. 21.

6. T. Kenner, "Arterial blood pressure and its measurement," Basic Res. Cardiol. **83**, 107–121 (1988).

7. A.D. Nashif, D.I.G. Jones, and J.P. Henderson, *Vibration Damping* (Wiley, New York, 1985), pp. 122–123.

8. J.P. Den Hartog, *Mechanical Vibrations* (Dover, New York, 1985).

9. J.R. Cameron and J.G. Skofronick, *Medical Physics* (Wiley, New York, 1978), p. 166.

10. U. Dinnar, *Cardiovascular Fluid Dynamics.* (CRC, Boca Raton, FL, 1981), pp. 140–144.

11. G.N. Jager, N. Westerhof, and A. Noordergraaf, "Oscillatory flow impedance in electrical analog of arterial system: representation of sleeve effect and non-Newtonian properties of blood," Circ. Res. **16**, 121–133 (1965).

12. M.G. Taylor, "Wave-travel in a non-uniform transmission line, in relation to pulses in arteries," Phys. Med. Biol. **10**, 539–550 (1965).

13. G. Austin, "Equation for model intracranial aneurysm with consideration of small dissipation term," Math. Biosci. **22**, 277–291 (1974).

14. W.J. Powers, "Cerebral hemodynamics in ischemic cerebrovascular disease," Ann. Neurol. **29**, 231–240 (1991).

15. H.A. Kontos, E.P. Wei, R.M. Navari, J.E. Levasseur, W.I. Rosenblum, and J.L. Patterson, Jr., "Responses of cerebral arteries and arterioles to acute hypotension and hypertension," Am. J. Physiol. **234**, H371–H383 (1978).

16. F.P. Tiecks, A.M. Lam, R. Aaslid, and D.W. Newell, "Comparison of static and dynamic cerebral autoregulation measurements," Stroke **26**, 1014–1019 (1995).

17. R. Aaslid, K.-F. Lindegaard, W. Sorteberg, and H. Nornes, "Cerebral autoregulation dynamics in humans," Stroke **20**, 45–52 (1989).

18. M.E. Clark, J.D. Martin, R.A. Wenglarz, W.A. Himwich, and F.M. Knapp, "Engineering analysis of the hemodynamics of the circle of Willis," Arch. Neurol. **13**, 173–182 (1965).

19. M.E. Clark, W.A. Himwich, and J.D. Martin, "A comparative examination of cerebral circulation models," J. Neurosurg. **29**, 484–494 (1968).

20. W.A. Himwich, F.M. Knapp, R.A. Wenglarz, J.D. Martin, and M.E. Clark, "The circle of Willis as simulated by an engineering model," Arch. Neurol. **13**, 164–172 (1965).

21. W.A. Himwich, and M.E. Clark, "Cerebral blood flow comparisons between model and prototype," J. Appl. Physiol. **31**, 873–879 (1971).

22. N. Avman and E.A. Bering, "A plastic model for the study of pressure changes in the circle of Willis and major cerebral arteries following arterial occlusion," J. Neurosurg. **18**, 361–365 (1961).

23. M.A. Helal, "Derivation of closed-form expression for the cerebral circulation models," Comput. Med. Biol. **24**, 103–118 (1994).

24. K.L. Zierler, "Theoretical basis of indicator-dilution methods for measuring flow and volume," Circ. Res. **10**, 393–407 (1962).

25. K.L. Zierler, "Theory of use of indicators to measure blood flow and extracellular volume and calculation of transcapillary movement of tracers," Circ. Res. **12**, 464–471 (1963).

26. N.A. Lassen, D.H. Ingvar, and E.Skinhøj, "Brain function and blood flow," Sci. Am. **239**, 62–71 (1978).

27. S.S. Kety and C.E. Schmidt, "The nitrous oxide method for the quantitative determination of cerebral blood flow in man: Theory, procedure, and normal values," J. Clin. Invest. **27**, 476–483 (1948).

28. H. Yonas, D.W. Johnson, and R.R. Pindzola, "Xenon-enhanced CT of cerebral blood flow," Sci. Am. Sci. Med. **2**, 58–67 (1995).

29. L. Hacein-Bey, R. Nour, J. Pile-Spellman, R. Van Heertum, P.D. Esser, and W.L. Young. "Adaptive changes of autoregulation in chronic cerebral hypotension with arteriovenous malformations: An acetazolamide-enhanced single-photon emission CT study," Amer. J. Neuroradiol. **16**, 1865–1874 (1995).

30. H. Ito, M. Koyama, R. Goto, R. Kawashima, S. Ono, H. Atsumi, K. Ishii, and H. Fukuda, "Cerebral blood flow measurement with iodine-123-IMP SPECT, calibrated standard input function and venous blood sampling," J. Nucl. Med. **36**, 2339–2342 (1995).

31. B.L. Holman and M.D. Devous, Sr., "Functional brain SPECT: The emergence of a powerful clinical method," J. Nucl. Med. **33**, 1888–1904 (1992).

32. J.C. Masdeu, L.M. Brass, B.L. Holman, and M.J. Kushner, "Brain single-photon emission computed tomography," Neurology **44**, 1970–1977 (1994).

33. P.H. Altrocchi, M. Brin, J.H. Ferguson, M.L. Goldstein, P.B. Gorelick, D.F. Hanley, D.J. Lange, M.R. Nuwer, and S. van den Noort, "Assessment of brain SPECT. Report of the Therapeutics and Technology Assessment Subcommittee of the American Academy of Neurology," Neurology **46**, 278–285 (1996).

34. M.M. Ter-Pogossian, M.E. Raichle, and B.E. Sobel, "Positron-emission tomography," Sci. Am. **243**, 170–181 (1980).

35. J.M. Hoffman and R.E. Coleman, "Perfusion quantitation using positron emission tomography," Invest Radiol. **27**, S22–S26 (1992).

36. R.J. Smith, L. Shao, R. Freifelder, and J.S. Karp, "Quantitative measurements of cerebral blood flow in volume imaging PET scanners," IEEE Trans. Nucl. Sci. **42**, 1018–1023 (1995).

37. K.T. Lucke, M.E. Kerr, and G.I. Chovanes, "Continuous bedside cerebral blood flow monitoring," J. Neurosci. Nurs. **27**, 164–173 (1995).

38. B.G. Harvey, *Introduction to Nuclear Physics and Chemistry.* (Prentice-Hall, Englewood Cliffs, NJ, 1969), p. 108.

39. D.R. Pickens, "Perfusion/diffusion quantitation with magnetic resonance imaging," Invest. Radiol. **27**, S12–S17 (1992).

40. D. Le Bihan, "Theoretical principles of perfusion imaging: Applications to magnetic resonance imaging," Invest. Radiol. **27**, S6–S11 (1992).

41. D.R. Enzmann, M.R. Ross, M.P. Marks, and N.J. Pelc, "Blood flow in major cerebral arteries measured by phase-contrast cine MR," Amer. J. Neuroradiol. **15**, 123–129 (1994).

42. M.E. Moseley, A. de Crespigny, and D.M. Spielman, "Magnetic resonance imaging of human brain function," Surg. Neurol. **45**, 385–391 (1996).

43. F. Ståhlberg, A. Ericsson, B. Nordell, C. Thomsen, O. Henriksen, and B.R.R. Persson. "MR imaging, flow and motion," Acta Radiol. **33**, 179–200 (1992).

44. R.G. Gould, "Perfusion quantitation by ultrafast computed tomography," Invest. Radiol. **27**, S18–S21 (1992).

45. M. Schoning, J. Walter, and P. Scheel, "Estimation of cerebral blood flow through color duplex sonography of the carotid and vertebral arteries in healthy adults," Stroke **25**, 17–22 (1994).

3.10   Problems

3.1. Why is the pulmonary circulation normally under low pressure?

3.2. Assuming that the heart ejects 80 cm^3 of blood under a systemic pressure of 150 mm Hg, estimate the work done on the left ventricle.

3.3. Compare the pressure distribution from systole to diastole between rigid and elastic tubes.

3.4. How can the veins sufficiently drive large volumes of blood to the heart under small pressures of approximately 10 mm Hg?

3.5. Consider a single tube of cross-sectional area A and assemblies of three smaller tubes connected in parallel and in series, each with a cross-sectional area $A/3$. How does the resistance of the single tube compare with the three smaller tubes connected in series and in parallel?

4
Hemodynamics: The Physics of Blood Flow

4.1 Introduction

Chapter 3 presented a global qualitative perspective of the anatomical and physiological components of the human circulatory system. That chapter was intended to provide the reader with a biophysical foundation for the structure, function, and demands of the blood vessels comprising the circulatory system, particularly those connected directly to the brain and cerebral circulation. The premise of this chapter represents, in effect, a continuation of the text presented in Chap. 3. The purpose of this chapter is to extend this perspective to a more localized one and to describe the dynamics of blood flow or hemodynamics. The role and overall importance of the vascular system and the dynamic interrelationships with the flowing blood are critical in all developmental aspects of cerebrovascular disease and will form the basis for discussion of topics in this chapter. However, it is the dynamics of the blood in combination with the mechanics of the blood vessels that dictate the onset and progression of cerebrovascular diseases.

Before we begin our discussion of hemodynamics or the physics of blood flow, the reader should first be introduced to descriptive terminology of a fluid in motion and its consequent implications into the biophysical interactions observed in hemodynamics. As can be seen from previous chapters, the physical behavior of a solid particle can be represented and understood easily because it constitutes a single entity of sufficient size that we can visualize such behavior as well. Extension of the same observations become more complex when dealing with fluids since we are, in effect, dealing with an ensemble of "virtual" particles that cannot be visualized. For example, the physics involved in dropping a stone from a building can be quantitated readily by enforcing the laws of kinematics or particle motion which are derived directly from Newton's laws of mechanics. A similar analysis of a bucket of water dropped from the same building cannot be performed with the same kinematic equations but can be quantitated by using Newton's laws as the basis for deriving appropriate equations.

The term *fluid is* used to describe an object or substance that must be in motion to resist externally applied forces or stresses. On a slightly windy day, a car, or any solid mass with significant weight for that matter, can withstand the forces exerted on it by the wind and remain stationary. The same could not be said for water flowing from a garden hose. Water flowing from the hose at a moderate rate will be sprayed in all directions. How do the forces from the wind act on the fluid and what are the mechanisms behind the behavior of the fluid? It should be noted that, although one tends to think of fluid primarily as liquids, fluids also describe the behavior of gases. The diference between liquids and gases will become apparent in defining the physical properties of a fluid.

4.2   Fluid Mechanics and Dynamics

In this section, the concept of a fluid is discussed in terms of a physical system. Let us consider the example of a glass of water defined by its density or mass per unit volume. A fluid contained within a defined volume such as a glass of water exerts a net force or pressure against the glass (body force), while the glass is exerting an equal yet opposite force against the water (surface force). It is also apparent that, since equal yet opposite forces are acting between the water and the glass, the water is in equilibrium, the net force of the system (glass and water) is equal to zero. Let us now assume that the water is contained within an open cylindrical tube. If one of the two forces mentioned previously exceeds the other in magnitude or if an external force acts on the fluid such as a pressure gradient applied between the ends of the tube, the fluid is set in motion and assumes an acceleration. The motion of the fluid caused by either the excess or external forces induces yet another force (viscous force) dependent on the property of viscosity unique to every fluid.

4.2.1   *Dynamic Characteristics of a Fluid*

The fluid behaves according to an assortment of various mechanisms depending on the fluid properties, the forces exerted on the fluid and the resultant motion, and the boundary conditions of the vessel geometry. The complexity of fluid motion and dynamics arises from the dynamic characteristics and physical properties of a fluid. The physical properties that characterize the nature and behavior of fluids are viscosity, laminar versus turbulent flow, steady versus pulsatile flow, compressible versus incompressible flow, and Newtonian versus non-Newtonian fluid flow.

4.2.1.1 Viscosity

Viscosity of a fluid is an inherent property that exerts a resistance or frictional force against the fluid motion in response to shear stress. The concept of viscosity can be illustrated through the comparison of two different types of fluid, namely, water and syrup. From common everyday experiences, it is known that syrup feels thicker and pours at a rate different from that of water. This is a simple demonstration of fluid viscosity. Syrup possesses a larger viscosity than that of water. Viscosity is, in essence, a proportionality constant between the shear stress exerted against the wall of the vessel and the rate of shear and is described according to Newton's law of fluid mechanics:

$$\text{Shear stress } (\tau) \propto \text{Shear rate (SR)}$$

or

$$\tau = \mu(\text{SR})$$

where μ is the viscosity coefficient. Solving for μ yields the following definition for viscosity:

$$\mu = \frac{\tau}{\text{SR}}.$$

The shear rate SR is the velocity gradient calculated along the radial axis (Fig. 4.1) and is defined mathematically by

$$\text{SR} = -\frac{dv}{dr},$$

where r is the radial coordinate of the cylindrical vessel and the negative sign accounts for the opposing direction exerted by the viscous force. Substituting in the expression for viscosity,

$$\mu = -\frac{\tau}{\dfrac{dv}{dr}}.$$

From the above relation, it can be seen that, for increasing velocity gradient,

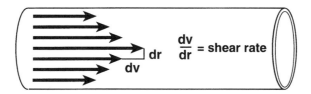

FIGURE 4.1. The layers or laminae characteristic of laminar flow represent the distribution of velocity along the cross-sectional diameter of the tube. The change in velocity represented by any two layers with respect to their corresponding position along the radial axis is the shear rate.

TABLE 4.1. Coefficients of viscosity for common fluids at room temperature.

Fluid	Coefficient of viscosity (cP)
Water	1.0
Olive oil	84.0
Ethyl alcohol	1.19
Methyl alcohol	0.59
Glycerine	830
Blood	3.5

the influence of viscous forces becomes minimal. This observation implies that viscosity plays a relatively minor flow in cases of blood flow through the aorta and larger vessels because of the larger velocity gradient. On the other hand, for the smaller vessels, particularly the capillary vessels, the viscous forces play a critical role in flow dynamics. Viscosity is measured in terms of N s/m^2 in SI units. Typical values of viscosity for several common fluids are listed in Table 4.1.

4.2.1.2   Laminar versus Turbulent Flow

Consider, for example, a garden hose attached to a tap. As one turns on the tap just enough to propel water through the faucet, the water exits the hose in a constant stream. As the water travels through the tubular hose, it assumes a parabolic profile where the velocity is at its maximum at the center of the profile and minimum toward the inner wall of the hose. The parabolic profile is visualized typically as continuous series of laminae or layers stacked upon one another. If we return to the previously established concept of fluid particles, the particles travel in continuous and orderly paths. Fluid flow exhibiting this type of velocity profile is termed *laminar flow*.

Returning to the garden hose, if the tap were turned toward its maximum level, the substantially increased pressure gradient would force the water to be expelled from the hose at a high velocity and in a random and disorderly fashion. This type of fluid flow is termed *turbulent flow*. In turbulent flow, the fluid particles fluctuate between the previously ordered laminae, causing mixing of the fluid.[1] The mixing and interacting of the fluid particles between layers cause a random motion and random transformation of energy into an acceleration of particles, deceleration of particles, and a dissipation into heat. This direct translation of velocity energy to heat energy is due to inertial forces and must be distinguished from viscous forces that cause the translation of heat energy from velocity energy through the viscous drag between adjacent fluid elements.[2] In turbulent flow, the velocity profile is much blunter in the central portion of the tube and much sharper at the wall. In turbulent flow, the large velocities experienced by the turbulent fluid are more evenly distributed near the center of the vessel. Shear strain rate at the tube wall is much higher for turbulent flow as compared to laminar flow of the same volumetric flow rate.[3]

The condition upon which flow becomes turbulent is dictated by the dimensionless parameter, Reynold's number Re. Reynold's number is defined as the ratio of inertial force to viscous force exerted by a given fluid or

$$ \mathrm{Re} = \frac{\text{(inertial force)}}{\text{(viscous force)}} = \frac{\rho v d}{\mu}, $$

where ρ is fluid density, v is the average velocity, d is the tube diameter, and μ is fluid viscosity. As a rule of thumb, fluid flow within a rigid tube is laminar for values of the Reynold's number less than 1200. In the human circulation, maximum Reynold's numbers over one cardiac cycle range from ≈ 6000 to $<10^{-3}$, in transport from the heart to the microcirculation.[4]

4.2.1.3   Steady versus Pulsatile Flow

Returning to the previous example of flow through a garden hose, water exiting the hose is subject to a constant pressure gradient and is consequently characteristic of steady flow. In most applications of fluid flow, particularly in applications pertaining to physiology, fluid does not flow in a steady stream but is subject to an oscillating pressure gradient resulting in pulsatile flow. Whether fluid flow is steady or pulsatile depends solely on the propagating or driving force initiating flow.

4.2.1.4   Compressible versus Incompressible Flow

Fluids are classified according to the physical behavior of both gases and liquids. The density of gases changes as a result of temperature and pressure changes. Thus, the flow of gases is considered compressible. The density of liquids, on the other hand, is relatively unaffected by changes in temperature and pressure. Consequently, the flow of liquids is considered incompressible.

4.2.1.5   Newtonian versus Non-Newtonian Fluid Flow

Let us return to the definition of viscosity presented in Sec. 4.2.1.1. Fluids behave according to the relation dictated by

$$ \tau = \mu(\mathrm{SR}), $$

where τ is the shear stress, μ is the fluid viscosity, and SR is strain rate. The linear relationship between the shear stress and shear rate characterizes a Newtonian fluid. However, for most fluids, shear stress is related in a nonlinear fashion with shear rate. This defines a non-Newtonian fluid.[5] In addition to the nonlinear relation exhibited by fluids, another scenario involves the dependence of shear stress not only on shear rate but normal strain as well. This is characteristic of a viscoelastic substance also considered a non-Newtonian fluid. Examples of common fluids that exhibit non-Newtonian behavior include ketchup, syrup, blood, paint, grease, pastes, inks, honey, and toothpaste. Representative examples of Newtonian and non-Newtonian fluids are summarized in Fig. 4.2.

FLUID	SHEAR STRESS	PHYSICAL BEHAVIOR
Ideal	$\tau = 0$	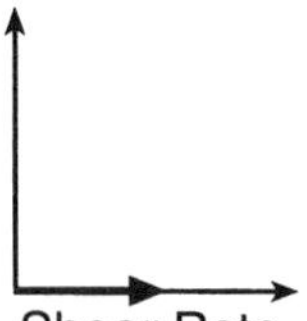 Shear Stress vs Shear Rate
Newtonian	$\tau = \mu \dfrac{dv}{dy}$	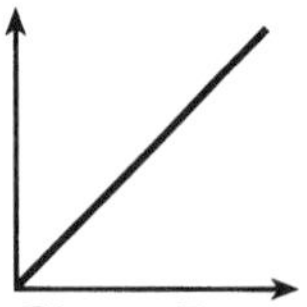 Shear Stress vs Shear Rate
Non-Newtonian	$\tau = \mu \left(\dfrac{dv}{dy}\right)^{n}$	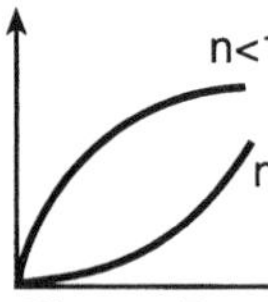 Shear Stress vs Shear Rate, n<1.0, n>1.0
Ideal Plastic	$\tau = \tau_y + \mu \left(\dfrac{dv}{dy}\right)$	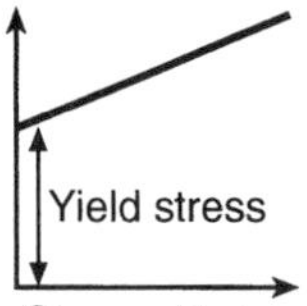 Shear Stress vs Shear Rate, Yield stress
Viscoelastic	$\tau + \left(\dfrac{\mu}{\lambda}\right)\dot{\tau} = \mu \left(\dfrac{dv}{dy}\right)$	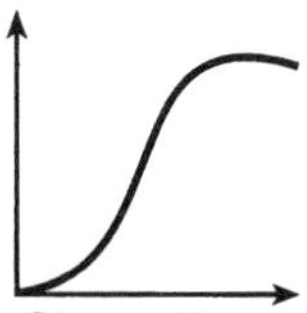 Shear Stress vs Shear Rate
τ_y = **yield stress** μ = **coefficient of viscosity**		λ = **rigidity modulus** n = **constant**

FIGURE 4.2. A description of the physical properties of Newtonian and non-Newtonian fluids.

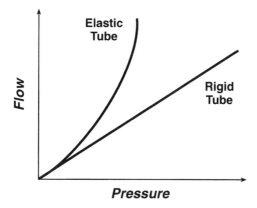

FIGURE 4.3. Schematic diagram illustrating the effects of pressure versus flow for a rigid and elastic tube.

4.2.1.6   Vessel Structure

Although not directly related to the physical properties of the fluid, the vessel geometry directly influences the dynamics of the fluid in motion. The first factor is the vessel elasticity. In many cases, particularly in the introduction of biophysical phenomena regarding fluid flow, one begins with fluid flow through a uniformly rigid tube. However, human blood vessels not only are elastic but are viscoelastic, exhibiting properties similar to both solids and fluids (Fig. 4.3). Another factor of vessel structure is vascular taper or a slight but progressive decrease in vessel diameter as the vessels become located farther from the heart. Although the angular decline or longitudinal changes in arterial diameter between distant segments of a single artery range from 0.4° to 1.0°,[6] abnormal hemodynamic conditions such as systemic hypertension could lower the Reynold's number, increasing the tendency for turbulence in an otherwise normal vessel. The final factor of vessel structure is vascular tethering. Blood vessels positioned in the human body are not isolated or free-standing but are tethered or permanently attached to adjacent tissue that acts as an additional external load on the structural wall of the vessel.

4.2.2   Equation of Continuity

As we consider the case of simple flow through a rigid tube, it is safe to assume that the total volume of fluid entering the tube will be equal to that exiting the tube. Furthermore, flow measured at one point along the tube will be equal to the flow at another point along the tube distal to the original point, regardless of the cross-sectional area of the tube at each point. This is due primarily to the fact that the fluid is incompressible, i.e., the fluid density is constant. Although energy from the blood flow is dissipated in the form of shear stress against the vessel wall, this energy loss is negligible,

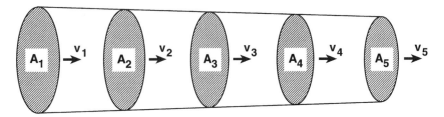

$$A_1 v_1 = A_2 v_2 = A_3 v_3 = A_4 v_4 = A_5 v_5$$

FIGURE 4.4. The equation of continuity, based on the conservation of mass, states that the product of velocity and cross-sectional area is equal at every point along a blood vessel, regardless of the degree of taper.

especially in large-sized tubes. This, in effect, is an illustration of the conservation of mass.

One example of the continuity equation relates to the flow in a blood vessel (Fig. 4.4). Although the geometry of human blood vessels can be approximated by an elastic cylinder, it also tapers or steadily decreases in size. The continuity equation guarantees that flow is equal at any point along the tapered tube, regardless of the degree of tapering. Thus, the relation between the flow at any two points within the tapered vessel is dependent on the cross-sectional area A of the vessel at the points of interest and their corresponding velocities v, which are given according to

$$A_1 v_1 = A_2 v_2$$

or

$$\frac{A_1}{A_2} = \frac{v_2}{v_1}.$$

One could also incorporate the length of the vessel segment, L, through which fluid flow velocity is measured by the same principle:

$$A_1 L_1 = A_2 L_2$$

or

$$\frac{A_1}{A_2} = \frac{L_2}{L_1}.$$

As another specific example of the continuity equation with applications to cerebrovascular disease, a major junction in the cerebrovasculature is the internal carotid artery that bifurcates or branches into two other smaller arteries supplying the brain with adequate blood circulation (Fig. 4.5). Although it may be desirable to make flow measurements within any vessel comprising the cerebral circulation, it might not be possible due to size limi-

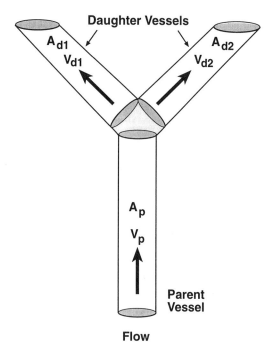

Flow

FIGURE 4.5. Schematic diagram illustrating an application of the continuity equation to an arterial bifurcation.

tations or constraints of the flowmeter and the overall condition of the patient. Thus, the ideal case would be to make flow measurements in the larger arteries and reasonably speculate about blood flow in the branching arteries based on this single blood flow measurement and caliber size measurements of the branching arteries assessed either from digital subtraction angiography or other three-dimensional (3-D) imaging modalities such as computed tomography (CT) or magnetic resonance (MR) angiography. Such approximations of blood flow within branching arteries are possible through the equation of continuity given as

(Flow in parent artery) = (Flow in branching or daughter arteries),

$$A_p \mathbf{v}_p = A_{d1} \mathbf{v}_{d1} + A_{d2} \mathbf{v}_{d2},$$

where A is the cross-sectional area of the vessel and $\mathbf{v}$ is the blood flow velocity. Obviously, since we have one equation and two unknowns, it is not possible to know exactly what the blood flow is in each daughter artery. It is possible, however, to make reasonable approximations based on the total flow from the parent artery by dividing the entire equation by the quantity

$$A_p v_p$$

to yield relative ratios of blood flow in the daughter arteries. In this case and as with all problems in hemodynamics, the density of blood is assumed to be constant and can thus be eliminated from the continuity equation.

The equation of continuity, as stated earlier, invokes the conservation of mass to fluid dynamics according to the following:

$$\frac{\partial \rho}{\partial t} + \nabla \cdot (\rho \mathbf{v}) = 0$$

where ρ is the density of the fluid, t is time, and $\mathbf{v}$ represents the velocity of a volume element of the fluid. This equation can be simplified significantly by making suitable assumptions. One assumption is steady flow. Steady flow implies no dependence on time or

$$\frac{\partial \rho}{\partial t} = 0.$$

The continuity equation then reduces to

$$\nabla \cdot (\rho \mathbf{v}) = 0.$$

Since the fluid under consideration is of a liquid state of matter, the density of the moving fluid is considered constant and thus incompressible. Since the fluid density is constant, it can be factored out of the revised continuity equation, which now simplifies to

$$\nabla \cdot \mathbf{v} = 0.$$

The continuity equation, in essence, states that the divergence of the velocity is equal to 0. From the discussion on vector analysis in Chap. 1, the continuity equation can be expanded in terms of a vector, $\mathbf{v}$, according to the following coordinate systems.

Rectangular coordinate system:

$$\nabla \cdot \mathbf{v} = \frac{\partial V_x}{\partial x} + \frac{\partial V_y}{\partial y} + \frac{\partial V_z}{\partial z} = 0.$$

Cylindrical coordinate system:

$$\nabla \cdot \mathbf{v} = \frac{1}{r}\frac{\partial (r V_r)}{\partial r} + \frac{1}{r}\frac{\partial V_\theta}{\partial \theta} + \frac{\partial V_z}{\partial z} = 0.$$

Spherical coordinate system:

$$\nabla \cdot \mathbf{v} = \frac{1}{r^2}\frac{\partial (r^2 V_r)}{\partial r} + \frac{1}{r \sin \theta}\frac{\partial (V_\theta \sin \theta)}{\partial \theta} + \frac{1}{r \sin \theta}\frac{\partial V_\phi}{\partial \phi} = 0.$$

4.2.3   Streamlines

A particular application of the gradient, first introduced in Chap. 1, which will become physically relevant in the description of fluid dynamics is that of a potential acting on a particle. Consider a particle in a field or an enclosed region within close proximity of other particles that exert a direct influence on the physical behavior of the particular particle. The concept of particle motion in a field is used commonly to describe the motion of a fluid. The state of the fluid prohibits one from characterizing the physical properties of the fluid not as a single particle, as one could easily do with a solid, but as an ensemble of "fluid" particles. Therefore, any physical equation describing fluid dynamics relates to the behavior of a single "fluid" particle.

The paths of these particles, which can be visualized by injecting contrast medium or dye into the flowing fluid, are well defined and can be characterized mathematically by streamlines. Streamlines are defined by two specific types of mathematical functions known as a velocity potential and a stream function. The purpose of these two functions is to locally define the magnitude and direction of a fluid particle along the curves presented by the two functions. Let us assume for the moment that we are dealing with two-dimensional fluid flow. The velocity of the fluid particle is described by the negative gradient of a velocity potential function, which, in essence, represents a scalar function, ϕ:

$$\mathbf{v} = -\nabla\phi,$$

where the components of the fluid velocity are

$$v_x = -\frac{\partial\phi}{\partial x}, \quad v_y = -\frac{\partial\phi}{\partial y}.$$

Since the fluid is still assumed to be incompressible, the velocity must satisfy the continuity equation:

$$\nabla \cdot \mathbf{v} = \nabla \cdot (-\nabla\phi) = -\nabla^2\phi = 0.$$

As you may recall from Chap. 1, this is Laplace's equation. Now that the velocity potential function has been introduced, let us direct our attention to the stream function. The stream function, ψ, represents a family of constant curves that are also related to the velocity of the fluid particle by a vector operation:

$$v_x = -\frac{\partial\psi}{\partial y}, \quad v_y = \frac{\partial\psi}{\partial x}.$$

The general equation describing a streamline in two-dimensional flow is

$$v_y\, dx - v_x\, dy = 0.$$

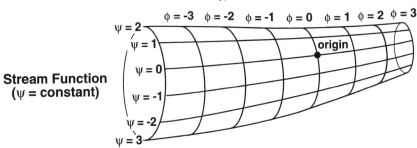

FIGURE 4.6. The position of a fluid particle can be described mathematically from the grid presented by the stream function and velocity potential function.

Substituting the individual velocity components into the above expression,

$$\frac{\partial \psi}{\partial x} dx + \frac{\partial \psi}{\partial y} dy = d\psi = 0.$$

Since the differential of ψ is 0, then ψ must be constant along a streamline. The next step in this process is to relate the two quantities ϕ and ψ. Since each function has been linked to the velocity components v_x and v_y, we can, in effect, relate the differentials corresponding to the two velocity components so that

$$v_x = -\frac{\partial \phi}{\partial x} = -\frac{\partial \psi}{\partial y}, \quad v_y = -\frac{\partial \phi}{\partial y} = \frac{\partial \psi}{\partial x}.$$

The purpose of stressing the fact that these functions satisfy the vector identities of divergence and Laplace's equation is to demonstrate that these functions are orthogonal or run perpendicular to each other. Thus, the overall purpose of implementing such concepts as the stream function and velocity potential function was to develop an idealized grid over the vessel geometry using the intersection points of these two functions to represent a localized region of the vessel. An example of the mapped grid lines using these two functions is displayed in Fig. 4.6.

4.3   Properties of Blood

Up to this point, discussions of fluid flow were made in reference to a generalized fluid that, in most cases, could be understood and investigated directly according to the fluid dynamic principles, interactions, and phenomena described in this book. However, since we are dealing with the circulatory system and blood flow, the discussion of fluid dynamics will be devoted primarily to blood. Blood is a unique fluid possessing specific components and properties necessary to perform the multitude of life-sustaining tasks, making it extremely difficult to characterize the fluid biophysically and experimentally. Blood is a complex fluid consisting of a fluid component (plasma 55%) and a solid component (blood cells 45%) composed of various particulates. The cells are heavier than the plasma component, making it easier to delineate and visualize these two components upon centrifugation. The physical density of blood, ρ, is 1.056 g/ml, and the viscosity is 3.5 cP (centipoise).

As stated previously, fluids are characterized typically as either Newtonian or non-Newtonian. The viscosity of a Newtonian fluid remains constant during flow and exhibits a linear relationship with shear stress. With respect to blood, its viscosity is not constant and changes with velocity, causing a variable shear stress along the artery wall. Thus, blood is a non-Newtonian fluid. In fact, blood behaves similar to a viscoelastic fluid. However, the non-Newtonian character of blood becomes significant in only the smallest blood vessels, such as the capillaries and the arterioles. In the larger blood vessels such as the aorta and major arteries, blood behaves more like a Newtonian fluid.

4.4   Hemodynamics

Hemodynamics involves the qualitative and quantitative study of the forces generated by and exerted upon the blood during transit. Blood is continually pumped directly from the heart into the arteries under systolic pressures ranging from 70 to 140 mm Hg. The pulsatile pressure induces oscillatory motions along the arterial wall that adversely affects blood flow. The arteries are the vessels most susceptible to the physical impact produced by the hemodynamic forces and are thus the most likely candidates for aneurysm development. In order to understand the physical nature of hemodynamics, we first must characterize mathematically blood flow in terms of system parameters. Our system in this case is the blood vessel, which is approximated as a rigid, cylindrical tube of radius R and length L subjected to a constant external pressure gradient ΔP. A fluid of viscosity μ is being propelled through the tube as a result of the pressure gradient. Given this information, our mission (which we have decided to accept) is to determine the rate of fluid flow in terms of the system parameters R, L, ΔP, and, μ.

4.4.1   Steady Flow through a Rigid Tube

We begin, as we would given any physical system, with an analysis of the physical entities involved in the motion of the fluid through the tube. Referring to the diagram in Fig. 4.7, there exist two forces acting on the fluid in opposing directions. The first force, F_P, is the force exerted by the pressure

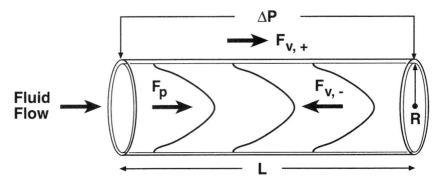

FIGURE 4.7. Schematic diagram illustrating the forces exerted by fluid flow through a rigid tube and the basis for the derivation of Poiseuille's law.

gradient and is defined as the pressure gradient multiplied by the area of the vessel over which it is exerted, i.e., the cross-sectional area of the vessel, πr^2, or

$$F_P = \Delta P(\pi r^2).$$

The other force acts in a direction opposite to that of F_P and is termed a viscous force, F_v. As it flows, the fluid, due to its viscosity, exerts a frictional or resistance force along the inner walls of the tube as a result of shear stresses. The total area affected by F_v is the internal surface area of the cylinder (SA_{cyl}) or

$$SA_{cyl} = 2\pi r L.$$

As it flows, the fluid exerts a viscous force against the fluid motion. Let us refer to this force as $F_{v,-}$ and define it as

$$F_{v,-} = 2\pi r L \mu \frac{dv}{dr}.$$

As mentioned previously, the shear stress is related to the shear rate according to:

$$\tau = \mu(\text{SR}) = \mu\left(-\frac{dv}{dr}\right).$$

The negative sign will be dropped from here on as long as it is understood that the shear stress acts in an opposing direction to the shear rate. There is an opposing force to this viscous force exerted along the outer surface of the vessel wall in the direction of fluid motion. This force, $F_{v,+}$, is given as

$$F_{v,+} = 2\pi r L \mu \frac{dv}{dr} + \frac{\partial}{\partial r}\left(2\pi r L \mu \frac{dv}{dr}\right).$$

The net viscous force F_v is the difference between these two components of the viscous force:

$$F_v = F_{v,+} - F_{v,-} = \frac{\partial}{\partial r}\left(2\pi r L \mu \frac{\partial v}{\partial r}\right).$$

Equating the two forces F_P and F_v yields

$$F_P = F_v,$$

$$\Delta P(\pi r^2) = \frac{\partial}{\partial r}\left(2\pi r L \mu \frac{dv}{dr}\right).$$

We are now left with a differential equation in terms of all of the system variables and a variable from which we could obtain information concerning fluid flow, i.e., the velocity v. Integrating this differential equation twice and using the assumption that $v = 0$ at the wall of the vessel, i.e., $r = R$, to determine the integration constant yields the following expression for the velocity:

$$v = \frac{\Delta P(R^2 - r^2)}{4L\mu}.$$

From the above equation, the velocity of the fluid at any point along the cross-sectional radius of the tube, i.e. from $r = 0$ (center of the tube) to $r = R$ (inner wall of the tube). This expression is known as the velocity profile and assumes a parabolic form or geometry. Elementary analysis of the velocity profile reveals a maximum velocity v_{max} at the center of the tube,

$$v_{max} = \frac{\Delta P R^2}{4L\mu},$$

and is at a minimum, v_{min}, at the inner wall of the tube,

$$v_{min} = 0.$$

This, remember, is an expression for the blood flow velocity or displacement of fluid over time (cm/s). It is, however, more convenient in some clinical applications to know the volumetric blood flow rate Q or motion of a volume of blood over time (cm^3/s). It is possible to derive an expression for the volumetric blood flow rate from the velocity profile by

$$Q = \pi R_t^2 v,$$

$$Q = \frac{\pi \Delta P R_t^4}{8L\mu}.$$

This is known as Poiseuille's law for fluid flow through a rigid tube.

One can also derive a differential equation for the fluid velocity by returning to the point in the derivation of Poiseuille's law where the pressure and viscous forces were equated to give

$$\frac{\partial}{\partial r} r \frac{\partial v}{\partial r} = -\frac{r \Delta P}{\mu L},$$

where r is the radial coordinate. The left-hand side of the above equation can be expanded according to the differentiation properties of a product and the entire equation can be rewritten:

$$\frac{\partial^2 v}{\partial r^2} + \frac{1}{r}\frac{\partial v}{\partial r} + \frac{\Delta P}{\mu L} = 0.$$

This differential equation can be solved to reveal the exact same solution for velocity as was obtained earlier in the section.

Although Poiseuille's law is cited frequently in physiology texts for discussion of vascular blood flow, it should be mentioned that there exist several assumptions that limit the true applicability of Poiseuille's law to hemodynamics in human blood vessels.[7] These assumptions include laminar flow; a uniform, rigid, cylindrical tube; a Newtonian fluid; a constant pressure gradient; and an axisymmetric velocity profile. Regardless of the limitations due to the assumptions, Poiseuille's law for fluid flow is generally accepted by medical scientists and provides a scientific means to investigate qualitatively and, to a lesser extent, quantitatively, the influence of parameters such as vessel length, vessel radius, and pressure gradient.

4.4.2 Pulsatile Flow through a Rigid Tube

Probably the most critical of assumptions in the derivation of Poiseuille's law is that of constant pressure. Physiological circulation is maintained by a pulsatile, oscillatory pressure gradient and thus, in an attempt to more closely represent the human circulation, should be implemented into hemodynamic calculations. Fluid dynamics can be described appropriately by defining mathematical relations based on the physical properties of conservation of mass and momentum. To do so, we must begin with a set of differential equations for fluid flow that are dependent not only on spatial location, but also on time. One of the equations needed to simulate fluid dynamics is the continuity equation, which was described in detail earlier. The second set of equations required to simulate pulsatile fluid flow is the Navier–Stokes equation.

4.4.2.1 Navier–Stokes Equation

The Navier–Stokes equation is an extension of Newton's second law, which states that for a fluid particle in motion within a stationary coordinate system, the product of the mass and acceleration of the fluid particle is equal to the sum of all external forces acting on the particle. The external forces exerted on the fluid particle are (1) gravitational force, (2) pressure forces, and

(3) viscous forces due to the internal friction within the fluid and are written in equation form as

$$\frac{\partial \mathbf{v}}{\partial t} + (\mathbf{v} \cdot \nabla)\mathbf{v} = F_g - \frac{1}{\rho}\nabla P + \frac{\mu}{\rho}\nabla^2 \mathbf{v},$$

where F_g represents gravitational or body forces, ρ is fluid density, and μ is fluid viscosity. As we have done throughout the text, the individual components of the three coordinate systems for the Navier–Stokes equations will now be written.

Rectangular coordinate system:

$$\frac{\partial V_x}{\partial t} + V_x\frac{\partial V_x}{\partial x} + V_y\frac{\partial V_x}{\partial y} + V_z\frac{\partial V_x}{\partial z} = F_g + -\frac{1}{\rho}\frac{\partial P}{\partial x} + \mu\left(\frac{\partial^2 V_x}{\partial x^2} + \frac{\partial^2 V_x}{\partial y^2} + \frac{\partial^2 V_x}{\partial z^2}\right),$$

$$\frac{\partial V_y}{\partial t} + V_x\frac{\partial V_y}{\partial x} + V_y\frac{\partial V_y}{\partial y} + V_z\frac{\partial V_y}{\partial z} = F_g + -\frac{1}{\rho}\frac{\partial P}{\partial y} + \mu\left(\frac{\partial^2 V_y}{\partial x^2} + \frac{\partial^2 V_y}{\partial y^2} + \frac{\partial^2 V_y}{\partial z^2}\right),$$

$$\frac{\partial V_z}{\partial t} + V_x\frac{\partial V_z}{\partial x} + V_y\frac{\partial V_z}{\partial y} + V_z\frac{\partial V_z}{\partial z} = F_g + -\frac{1}{\rho}\frac{\partial P}{\partial z} + \mu\left(\frac{\partial^2 V_z}{\partial x^2} + \frac{\partial^2 V_z}{\partial y^2} + \frac{\partial^2 V_z}{\partial z^2}\right).$$

Cylindrical coordinate system:

$$\frac{\partial V_r}{\partial t} + V_r\frac{\partial V_r}{\partial r} + \frac{V_\phi}{r}\frac{\partial V_r}{\partial \phi} + V_z\frac{\partial V_r}{\partial z} - \frac{V_\phi^2}{r}$$

$$= F_g + -\frac{1}{\rho}\frac{\partial p}{\partial r} + \mu\left[\frac{1}{r}\frac{\partial}{\partial r}\left(r\frac{\partial V_r}{\partial r}\right) + \frac{1}{r^2}\frac{\partial^2 V_r}{\partial \phi^2} + \frac{\partial^2 V_r}{\partial z^2} - \frac{2}{r^2}\frac{\partial V_\phi}{\partial \phi} - \frac{V_r}{r^2}\right],$$

$$\frac{\partial V_\phi}{\partial t} + V_r\frac{\partial V_\phi}{\partial r} + \frac{V_\phi}{r}\frac{\partial V_\phi}{\partial \phi} + V_z\frac{\partial V_\phi}{\partial z} - \frac{V_r V_\phi}{r}$$

$$= F_g + -\frac{1}{\rho r}\frac{\partial p}{\partial \phi} + \mu\left[\frac{1}{r}\frac{\partial}{\partial r}\left(r\frac{\partial V_\phi}{\partial r}\right) + \frac{1}{r^2}\frac{\partial^2 V_\phi}{\partial \phi^2} + \frac{\partial^2 V_\phi}{\partial z^2} + \frac{2}{r^2}\frac{\partial V_r}{\partial \phi} - \frac{V_\phi}{r^2}\right],$$

$$\frac{\partial V_z}{\partial t} + V_r\frac{\partial V_z}{\partial r} + \frac{V_\phi}{r}\frac{\partial V_z}{\partial \phi} + V_z\frac{\partial V_z}{\partial z}$$

$$= F_g + -\frac{1}{\rho}\frac{\partial p}{\partial z} + \mu\left[\frac{1}{r}\frac{\partial}{\partial r}\left(r\frac{\partial V_z}{\partial r}\right) + \frac{1}{r^2}\frac{\partial^2 V_z}{\partial \phi^2} + \frac{\partial^2 V_z}{\partial z^2}\right].$$

Spherical coordinate system:

$$\frac{\partial V_r}{\partial t} + V_r\frac{\partial V_r}{\partial r} + \frac{V_\theta}{r}\frac{\partial V_r}{\partial \theta} + \frac{V_\phi}{r\sin\theta}\frac{\partial V_r}{\partial \phi} - \frac{V_\theta^2 + V_\phi^2}{r}$$

$$= F_g + -\frac{1}{\rho}\frac{\partial p}{\partial r} + \mu\left[\frac{1}{r^2}\frac{\partial}{\partial r}\left(r^2\frac{\partial V_r}{\partial r}\right) + \frac{1}{r^2\sin\theta}\frac{\partial}{\partial \theta}\left(\sin\theta\frac{\partial V_r}{\partial \theta}\right) + \frac{1}{r^2\sin^2\theta}\frac{\partial^2 V_r}{\partial \phi^2}\right.$$

$$\left. - \frac{2}{r^2\sin^2\theta}\frac{\partial(V_\theta\sin\theta)}{\partial \theta} - \frac{2}{r^2\sin^2\theta}\frac{\partial V_\phi}{\partial \phi} - \frac{2V_r}{r^2}\right],$$

$$\frac{\partial V_\theta}{\partial t} + V_r \frac{\partial V_\theta}{\partial r} + \frac{V_\theta}{r} \frac{\partial V_\theta}{\partial \theta} + \frac{V_\phi}{r \sin \theta} \frac{\partial V_\theta}{\partial \phi} + \frac{V_r V_\theta}{r} - \frac{V_\phi^2 \cot \theta}{r}$$

$$= F_g + -\frac{1}{\rho r} \frac{\partial p}{\partial \theta} + \mu \left[\frac{1}{r^2} \frac{\partial}{\partial r} \left(r^2 \frac{\partial V_\theta}{\partial r} \right) + \frac{1}{r^2 \sin \theta} \frac{\partial}{\partial \theta} \left(\sin \theta \frac{\partial V_\theta}{\partial \theta} \right) + \frac{1}{r^2 \sin^2 \theta} \frac{\partial^2 V_r}{\partial \phi^2} \right.$$

$$\left. - \frac{2 \cos \theta}{r^2 \sin^2 \theta} \frac{\partial V_\phi}{\partial \phi} + \frac{2}{r^2} \frac{\partial V_r}{\partial \theta} - \frac{V_\theta}{r^2 \sin^2 \theta} \right],$$

$$\frac{\partial V_\phi}{\partial t} + V_r \frac{\partial V_\phi}{\partial r} + \frac{V_\theta}{r} \frac{\partial V_\phi}{\partial \theta} + \frac{V_\phi}{r \sin \theta} \frac{\partial V_\phi}{\partial \phi} + \frac{V_r V_\phi}{r} + \frac{V_\theta V_\phi \cot \theta}{r}$$

$$= F_g + -\frac{1}{\rho r \sin \theta} \frac{\partial p}{\partial \phi} + \mu \left[\frac{1}{r^2} \frac{\partial}{\partial r} \left(r^2 \frac{\partial V_\phi}{\partial r} \right) + \frac{1}{r^2 \sin \theta} \frac{\partial}{\partial \theta} \left(\sin \theta \frac{\partial V_\phi}{\partial \theta} \right) \right.$$

$$\left. + \frac{1}{r^2 \sin^2 \theta} \frac{\partial^2 V_\phi}{\partial \phi^2} + \frac{2}{r^2 \sin \theta} \frac{\partial V_r}{\partial \phi} + \frac{2 \cos \theta}{r^2 \sin^2 \theta} \frac{\partial V_\theta}{\partial \phi} - \frac{V_\phi}{r^2 \sin^2 \theta} \right].$$

The viscous forces are proportional to the viscosity of the fluid and depend on the velocity gradient within the flow space. In most cases of fluid flow, the gravitational force is negligible and will be omitted in all expressions of the Navier–Stokes equations. For a fluid with small viscosity and for a fluid with small velocity variations, the viscous forces may be small and negligible. As mentioned previously, blood is a non-Newtonian fluid and becomes significant from a hemodynamic point of view only in the smallest blood vessels. As displayed above, the Navier–Stokes equations contain nonlinear terms that make it impossible to solve either analytically or numerically. Thus, for a given case of fluid flow, one must implement approximations with the ultimate goal of linearizing the Navier–Stokes equations. Let us assume for the moment that we would like to represent steady axisymmetric fluid flow through a rigid cylindrical tube. In our calculations, we would like to simplify Navier–Stokes equations by omitting the non-linear terms.[8] In Navier–Stokes equations, the viscous force term can be expanded utilizing vector identities:

$$\nabla^2 \mathbf{V} = \nabla(\nabla \cdot \mathbf{V}) - \nabla \times (\nabla \times \mathbf{V}).$$

The nonlinear term

$$\nabla \times (\nabla \times \mathbf{V})$$

is a term that would simplify calculations but we need to determine whether this term can be properly omitted. This can be shown through elementary vector analysis. The flow can be represented by a vector with a component in the z direction or

$$\mathbf{v} = v\mathbf{z}.$$

This vector is now implemented into the vector operation presented by the nonlinear term. We first begin by performing the vector operation contained within the parentheses in the previous equation for a cylindrical geometry:

$$\nabla \times \mathbf{v} = \frac{1}{r} \begin{vmatrix} \mathbf{r} & r\boldsymbol{\phi} & \mathbf{z} \\ \dfrac{\partial}{\partial r} & \dfrac{\partial}{\partial \phi} & \dfrac{\partial}{\partial z} \\ 0 & 0 & v \end{vmatrix} = \mathbf{r}\frac{\partial v}{\partial \phi} - \boldsymbol{\phi}\frac{\partial v}{\partial r}.$$

Using this result, the final vector operation can be performed in a similar fashion:

$$\nabla \times (\nabla \times \mathbf{v}) = \frac{1}{r} \begin{vmatrix} \mathbf{r} & r\boldsymbol{\phi} & \mathbf{z} \\ \dfrac{\partial}{\partial r} & \dfrac{\partial}{\partial \phi} & \dfrac{\partial}{\partial z} \\ \dfrac{\partial v}{\partial \phi} & -\dfrac{\partial v}{\partial r} & 0 \end{vmatrix} = \mathbf{r}\left[-\frac{\partial}{\partial z}\left(\frac{\partial v}{\partial r}\right)\right] + r\boldsymbol{\phi}\left[\frac{\partial}{\partial z}\left(\frac{\partial v}{\partial \phi}\right)\right]$$

$$+ \mathbf{z}\left[\frac{\partial}{\partial r}\left(-\frac{\partial v}{\partial r}\right) - \frac{\partial}{\partial \phi}\left(\frac{\partial v}{\partial \phi}\right)\right] = \mathbf{0}.$$

Since the velocity has only a z component, partial derivatives with respect to r and ϕ are zero. Thus, the total value of the combined vector operation is zero. With the appropriate initial and boundary conditions, the velocity and pressure from linearized Navier–Stokes equations can be determined at any time using advanced numerical technique such as finite-element analysis.[9]

Returning to the subject at hand, we wish to obtain an expression for the velocity of fluid driven through a rigid cylindrical tube by a pulsatile pressure gradient. The velocity equation used for this case is similar to that for steady flow with the exception of an added inertial term and now becomes[10]

$$\rho\frac{\partial v}{\partial t} = -\frac{\partial p}{\partial x} + \mu\left(\frac{\partial^2 v}{\partial r^2} + \frac{1}{r}\frac{\partial v}{\partial r}\right).$$

An oscillatory complex wave function is assigned to the pressure gradient as was done originally by Womersley[11]:

$$-\frac{\partial p}{\partial x} = Pe^{i\omega t},$$

where P is the amplitude of the pressure wave, i is a complex number, ω is the frequency of the periodic pressure wave, and t is time. Since the pressure is related to the velocity, the velocity is expressed in similar terms to the pressure wave function:

$$v = V_0(r)e^{i\omega t}.$$

Substituting the expressions for pressure and velocity into the differential equation for pulsatile flow,

$$\mu \frac{\partial^2 V_0}{\partial r^2} + \frac{\mu}{r} \frac{\partial V_0}{\partial r} - \rho V_0 i\omega = -P.$$

We wish to simplify this equation by moving $-P$ to the left-hand side of the equation, multiplying by r^2, and dividing by μ,

$$r^2 \frac{\partial^2 V_0}{\partial r^2} + r \frac{\partial V_0}{\partial r} - \frac{r^2}{\mu}(\rho V_0 i\omega - P) = 0.$$

This equation can be further simplified by defining another variable, σ,

$$\sigma = \rho i\omega V_0 - P,$$

and, upon substitution, the equation now becomes

$$r^2 \frac{\partial^2 \sigma}{\partial r^2} + r \frac{\partial \sigma}{\partial r} - \frac{\rho}{\mu}\sigma r^2 i\omega = 0,$$

which is of the form

$$x \frac{d^2 y}{\partial x^2} + \frac{\partial y}{\partial x} - i\beta\omega xy = 0,$$

where β is a constant. The purpose of such a simplification process is to manipulate the original equation into a form for which a known solution exists. The solution to the flow equation presented above is of the form:

$$y = CJ_0(i^{3/2} x \sqrt{\beta\omega}),$$

where J_0 is a Bessel function solution and C is a constant. Incorporating appropriate boundary conditions, the fluid flow velocity is of the form

$$v = \frac{P}{i\omega\rho}\left(1 - \frac{J_0\left(i^{3/2}\sqrt{\frac{\rho}{\mu}\omega R^2}\frac{r}{R}\right)}{J_0\left(i^{3/2}\sqrt{\frac{\rho}{\mu}\omega R^2}\right)}\right) e^{i\omega t},$$

and the corresponding volumetric blood flow rate is

$$Q = \frac{\pi R^4 P}{i\mu\alpha^2}\left(1 - \frac{2J_1(\alpha i^{3/2})}{\alpha i^{3/2} J_0(\alpha i^{3/2})}\right) e^{i\omega t},$$

where J_0 and J_1 represent Bessel functions of order 0 and 1, respectively, and α is a frequency parameter defined as

$$\alpha = R\sqrt{\frac{\rho\omega}{\mu}}.$$

The Bessel functions represent a family of solutions to the Bessel differential equation given by

$$x^2 \frac{d^2 y}{dx^2} + x \frac{dy}{dx} + (x^2 - n^2)y = 0, \quad n \geq 0.$$

Mathematical tables of the particular Bessel functions are typically consulted for particular cases of fluid flow (see Chap. 1 and Ref. 12).

Another issue in the case of pulsatile flow, which will be addressed briefly, is the propagation of pressure waves during flow. Assuming no reflections, the pressure gradient and flow for the case of pulsatile flow in a cylindrical tube are related, according to Ohm's law for alternating electric currents and oscillatory voltage sources, by the impedance Z:

$$Z_n = \frac{P_n(t)}{Q_n(t)},$$

where

$$P(t) = \sum_{n=0} P_n e^{i(\omega n t - \phi_n)}, \quad Q(t) = \sum_{n=0} Q_n e^{i(\omega n t - \theta_n)},$$

and n is the nth harmonic. However, at vascular points such as arterial bifurcations and vascular conditions presented by cerebrovascular diseases such as atherosclerotic stenoses (Chap. 5), the pressure waves reflect and propagate backwards. Thus, proper characterization of the pressure wave can be seen in the form

$$P(z, t) = P_b e^{i\omega t - \gamma z} + P_f e^{i\omega t + \gamma z},$$

where γ is a propagation constant and P_f and P_b are the amplitudes of the pressure waves in the forward and backward directions, respectively. For the elementary case of an elastic cylindrical vessel, the phase velocity c_p of the pressure waves is given by the Moens–Korteweg equation,[13]

$$c_p^2 = \frac{Eh}{2\rho R},$$

where E is the elastic modulus of the vessel, h is the wall thickness, ρ is the density of fluid (blood), and R is the vessel radius.

Comparison of the flow velocity and flow rate between the two cases of fluid flow presented (steady flow versus pulsatile flow) reveals a substantial increase in mathematical and physical complexity. Fluid flow has been investigated incorporating other factors such as elasticity of the cylindrical tube and non-Newtonian fluid have been investigated[14–17] but space limitations prohibit the proper discussion of these cases.

The Navier–Stokes equations form the basis for simulating fluid dynamics in complicated geometrical arrangements using computational methods and dedicated software programs. Computational methods represent the most widely used theoretical technique for the qualitative and quantitative investigation of fluid flow. Other techniques designed to study blood flow through blood vessels include mathematical models based on electrical network theory[18] and experimental *in vitro* models using contrast dyes for the visualization of fluid dynamics.[19-24] Each of these techniques play an important role in the study of blood flow, not only for blood vessels in the normal state, but pathophysiological states leading to the progression of cerebrovascular diseases.

4.5   Experimental Measurement Techniques of Vascular Blood Flow

Hemodynamic measurements are very important in the assessment of normal physiological conditions as well as in the development and presence of cerebrovascular diseases. Hemodynamic measurements, particularly blood flow rate and velocity, are integral to the diagnosis, management, and therapy of cerebrovascular diseases. The *in vivo* measurement of blood flow within arteries is important in the diagnosis of vascular disease. Instruments designed for such measurements are termed *flowmeters* and operate on various physical principles and interactions. Specific examples of flowmeters include electromagnetic flowmeters and ultrasonic Doppler probes. The ultrasonic and electromagnetic flowmeters, although very different in operation, are similar in their basic concept: they both operate by coupling a field through a vessel wall and measuring, external or internal to the vessel, a parameter that is dependent upon the velocity of flow.[25] Another method for the quantitative determination of blood flow involves the use of video-densitometric methods.

4.5.1   *Electromagnetic Flowmeters*

An electromagnetic flowmeter is an instrument used to quantitatively measure blood flow velocity through an artery. The flowmeter, shown in Fig. 4.8, consists of a magnetic core or ring clamped around an arterial segment that produces a magnetic field. The flowmeter is based on the principle that an electromotive force (emf) is induced in a conductor moving so as to cut the lines of force in a magnetic field. The induced emf is detected by signal electrodes and subsequently amplified. As the conductor or, in this case, blood, moves through the field in a direction perpendicular to its own axis and to the lines of force, then the potential difference in volts at the ends of the artery, in terms of the blood flow velocity, is given as

$$V = B\,dv \times 10^{-8}\,\text{V}$$

where B is the strength of the magnetic field (G, gauss), d is the internal

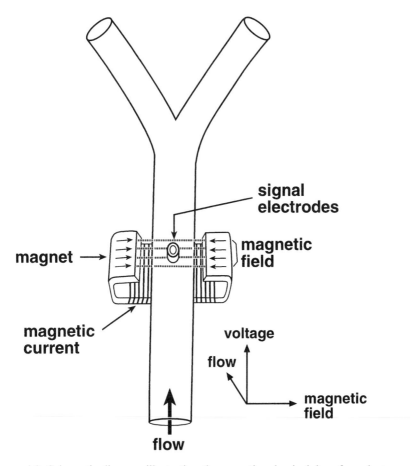

FIGURE 4.8. Schematic diagram illustrating the operational principles of an electro-magnetic flowmeter.

diameter of the artery (cm), and v is the average velocity of the flowing blood (cm/s). The volumetric blood flow rate Q is determined from the equation

$$Q = \frac{\pi d^2}{4} \frac{V}{Bd}.$$

This equation for flow is true provided three criteria are met: (1) the magnetic field is uniform, (2) the conductor moves in a plane at right angles to the magnetic field, and (3) the length of the conductor extends at right angles to both the magnetic field and direction of motion. Electromagnetic flowmeters have very good temporal resolution and are used for experimental studies of pulsatile flow, but they require direct exposure of a vessel, limiting clincial applications to measurements during surgery.

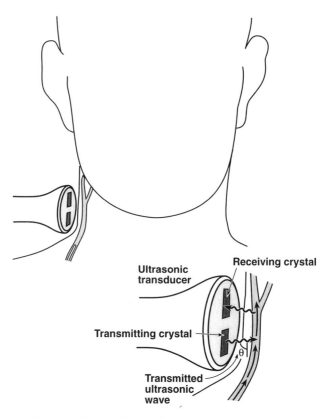

FIGURE 4.9. Schematic diagram illustrating the operational principles of an ultrasonic Doppler probe for flow measurements.

4.5.2   Ultrasonic Doppler Probe Flowmeter

The most popular and effective type of flowmeter is the ultrasonic probe flowmeter based on the Doppler effect. The working principles of the Doppler probe are described in Fig. 4.9. Blood flow is measured using an endovascular or external probe, consisting of a transducer with two crystals. One crystal produces ultrasonic waves that interact and reflect off of red blood cells before returning for detection by the other crystal. In brief, with the probe positioned over or in the vessel of interest, a continuous ultrasound beam of frequency f_0 impinges upon the blood cells and is scattered and reflected back toward the probe at a slightly lower Doppler-shifted frequency. The detected frequency is defined mathematically by

$$f_s = \frac{2f_0 V}{V_s} \cos\theta,$$

where V is the blood velocity, V_s is the velocity of sound in tissue ($=1550$ m/s),

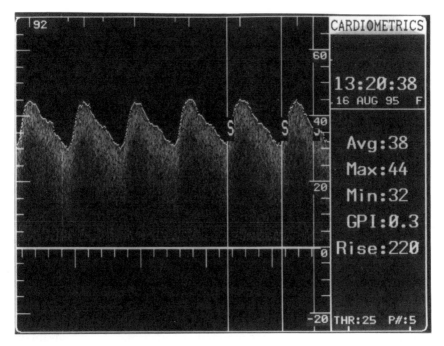

FIGURE 4.10. Flow wave form and quantitative flow values obtained from a Doppler flowmeter in the carotid artery of a laboratory swine.

and θ is the angle of incidence between the longitudinal flow axis and a line joining the two crystals of the ultrasound probe. The blood flow velocity V in the Doppler equation is determined from Poiseuille's law. Figure 4.10 shows blood flow measurements acquired from a Doppler flowmeter in the carotid artery of a laboratory swine. Endovascular flowmeters also exhibit good temporal resolution but require intravascular placement of the Doppler probe and proper positioning of the wire in the center of the vessel.

4.5.3   Videodensitometric Methods

Digital subtraction angiography is a medical imaging modality that employs rays and contrast agent to permit the visualization of blood vessels.[26,27] Since the contrast agent is opaque and mixes well with blood, one can easily track the contrast agent through successive angiographic image projections and obtain valuable information regarding the flow behavior and surrounding vascular arrangement. For the determination of blood flow velocity and flow rate from angiographic sequences, one is confronted with the task of first computing the time difference of the contrast bolus between two points along an arterial segment and then estimating the distance traversed by the bolus to arrive at rate or velocity information.

A variety of methods for the determination of blood flow velocity using videodensitometric techniques presently exist but can be grouped collectively according to two distinct subdivisions: time-of-flight of contrast bolus along a predetermined length of vessel[28-33] and the rate of dilution of contrast in the blood as a measure of flow rate.[34-37] In the former method, one obtains mass and density measurements of the bolus within a cross-sectional area of the vessel between two images within the sequence and uses the conservation of mass to obtain a mean flow velocity. The latter method involves integration of the distance–density curves over the sequence to yield relative values of bolus flow.

These methods have succeeded in producing typically accurate results of the blood flow rate but fail in calculations of blood flow velocity primarily since the curves give only average or relative values and not information on a pixel-by-pixel basis, which is mandatory for velocity measurements. In addition, these methods, based on the concept of measuring the differences of transit time of bolus flow within a fixed arterial segment, work well on constant flow but suffer greatly when applied to pulsatile flow. This makes it very hard for these methods to gain clinical acceptance, due to the pulsatility originating from the cardiac cycle. Finally, very high flow rates are extremely difficult to quantify with videodensitometric methods primarily due to the instability of the concentration gradient. This disadvantage makes these methods suspect in acquiring blood flow information in vessel pathologies such as intracranial aneurysms where turbulence or highly disordered flow is known to occur.[38]

Applications of videodensitometric methods for flow quantitation are limited by a host of factors commonly experienced during angiography including: (1) overlying anatomical vessels and structures; (2) variations in cerebrovascular geometry, including tapered, curved, sigmoidal or S-shaped, and tortuous vessels; (3) diffusion or streaming of contrast agent that depends on a number of parameters such as injection pressure, injection time, and blood pressure; (4) changes in image background due to x-ray flux; (5) changes in contrast agent concentration and injection site; (6) vessel views perpendicular to image view; (7) patient motion during scan or between scans; (8) frame rates <30 frames per second; (9) vessel size, length, and geometry; and (10) range of flow velocity >100 cm/s.

4.6   Hemodynamics of a Catheter

A catheter is a long, thin, flexible tube designed to give the physician vascular access to almost any point or region within the human body without having to surgically expose the region of interest. It is introduced into the body through a small incision, usually in the femoral artery of the leg, and navigated through the myriad of vessels with the use of x-ray fluoroscopy and angiography to the region or organ of interest. Several uses of a catheter include the delivery of contrast medium for angiographic visualization and delivery of embolic agents to effectively treat cerebrovascular diseases and for making quantitative and graphic hemodynamic measurements such as the pressure and blood flow velocity wave forms within the blood vessels.

The presence of a catheter, depending on its size, can increase vascular resistance, corresponding to flow disturbances.[39] Hemodynamic changes due to the intraarterial injection of contrast agent include: (1) reduction of blood flow due to the arterial puncture and catheter placement; (2) changes in arterial pressure distal to the site of injection; (3) change of flow pattern during the injections; and (4) changes in systemic blood pressure.[39,40] Hemodynamics of a fluid, i.e., contrast agent, can be described according to the previous description of blood flow through a rigid, cylindrical vessel. However, the fluid is forced through the catheter via an external force and is ejected in the form of a jet.[41,42] Jets represent the turbulent fluid flow exiting a constricted area such as the catheter tip or an arterial stenosis (Chap. 5). The fluid flow exiting the catheter deviates from the standard Poiseuille's formula and has been determined experimentally as[43]

$$F = \frac{\pi r^{25/7} P^{4/7}}{c^{4/7} \rho^{3/7} \eta^{1/7}},$$

where r is the radius of the catheter, η is the viscosity of the fluid, P is the pressure per unit length, c is a constant ($=0.0664$), and ρ is the density of the fluid. The role of catheters is instrumental in both the diagnosis and therapy of cerebrovascular diseases as will be evident in Chaps. 5–7.

4.7   Summary

Hemodynamic disturbances are implicated in the onset, development, diagnosis, and therapy of cerebrovascular diseases. This chapter summarized the important aspects of mathematical characterization of blood flow within a blood vessel as well as presenting theoretical and experimental methods for obtaining quantitative measurements of hemodynamics parameters such as pressure and flow. The next three chapters address the specific type of cerebrovascular disease: stroke (Chap. 5), intracranial aneurysms (Chap. 6), and arteriovenous malformations (Chap. 7). This chapter was written such that the information presented here and in the previous chapters will serve as an adequate basis for understanding the physics of cerebrovascular diseases.

4.8   References

1. S. Corrsin, "Turbulent flow," Am. Sci. **49**, 300–325 (1961).
2. E.O. Attinger, editor, *Pulsatile Blood Flow* (McGraw-Hill, New York, 1964).
3. Y.C. Fung, *Biodynamics: Circulation* (Springer, New York, 1984), Chap. 1.

4. V.T. Turitto and H.L. Goldsmith, "Rheology, transport, and thrombosis in the circulation," Loscalzo J, Creager MA, Dzau VJ (eds). *Vascular Medicine: A Textbook of Vascular Biology and Diseases*, edited by J. Loscalzo, M.A. Creager, and V.J. Dzau (Little, Brown, Boston, 1992), Chap. 5, 157–204.

5. R.V. Sharman. "Non-Newtonian liquids," Phys. Ed. **3**, 375–376 (1968).

6. W.R. Milnor, *Hemodynamics*, 2nd ed, (Williams & Wilkins, Baltimore, 1989), p. 42.

7. A. Roos, "Poiseuille's law and its limitations in vascular systems," Med. Thorac. **19**, 224–238 (1962).

8. G.B. Arfken, *Mathematical Methods for Physicists*, 3rd ed. (Academic, Orlando, 1985), p. 96.

9. J.E. Akin, *Finite Elements for Analysis and Design* (Academic, San Diego, 1994).

10. W.M. Phillips, "Modeling of flows in the circulatory system," Adv. Cardiovasc. Phys. 5, 26–48 (1983).

11. J.R. Womersley, "Method for the calculation of velocity, rate of flow and viscous drag in arteries when the pressure gradient is known," J. Physiol. **127**, 553–563 (1955).

12. N.W. McLachlan, *Bessel Functions for Engineers* (Oxford University Press, Oxford, 1941).

13. E.O. Attinger, "Structure and function of the peripheral circulation," in *Engineering Principles in Physiology*, edited by J.H.V. Brown and D.S. Gann (Academic, New York, 1973), p. 21.

14. R.H. Cox, "Comparison of linearized wave propagation models for arterial blood flow analysis," J. Biomech. **2**, 251–265 (1969).

15. H. Branson, "The flow of a viscous fluid in an elastic tube: A model of the femoral artery," Bull. Math. Biophys. **7**, 181–188 (1945).

16. J.P.W. Baaijens, A.A. van Steenhoven, and J.D. Janssen, "Numerical analysis of steady generalized newtonian blood flow in a 2D model of the carotid artery bifurcation," Biorheology **30**, 63–74 (1993).

17. K. Perktold, R.O. Peter, M. Resch, and G. Langs, "Pulsatile non-Newtonian blood flow in three-dimensional carotid bifurcation models: A numerical study of flow phenomena under different bifurcation angles," J. Biomed. Eng. **13**, 507–515 (1991).

18. A. Noordegraaf, "Hemodynamics," in *Biological Engineering*, edited by H.P. Schwan (McGraw-Hill, New York, 1969).

19. C.W. Kerber and C.B. Heilman, "Flow dynamics in the human carotid artery: I. preliminary observations using a transparent elastic model," Amer. J. Neuroradiol. **13**, 173–180 (1992).

20. R. Frayne, L.M. Gowman, D.W. Rickey, D.W. Holdsworth, P.A. Picot, M. Drangova, K.C. Chu, C.B. Caldwell, A. Fenster, and B.K. Rutt, "A geometrically accurate vascular phantom for comparative studies of x-ray, ultrasound, and magnetic resonance vascular imaging: construction and geometrical verification," Med. Phys. **20**, 415–425 (1993).

21. D.W. Liepsch and S.T. Moravec, "Pulsatile flow of non-Newtonian fluid in distensible models of human arteries," Biorheology **21**, 571–586 (1984).

22. T. Karino and M. Motomiya, "Flow visualization in isolated transparent natural blood vessels," Biorheology **20**, 119–127 (1983).

23. C.W. Kerber, C.B. Heilman, and P.H. Zanetti, "Transparent elastic arterial models I: A brief technical note," Biorheology **26**, 1041–1049 (1989).

24. C.W. Kerber and D. Liepsch, "Flow dynamics for radiologists. I. Basic principles of fluid flow," Amer. J. Neuroradiol. **15**, 1065–1075 (1994).

25. S. Yerushalmi and Y. Itzchak, "Angiographic methods for blood flow measurements," Med. Prog. Technol. **4**, 107–115 (1976).

26. D.C. Levin, R.M. Schapiro, L.M. Boxt, L. Dunham, D.P. Harrington, and D.L. Ergun, "Digital subtraction angiography: Principles and pitfalls of image improvement techniques," Amer. J. Roentgenol. **143**, 447–454 (1984).

27. D.P. Harrington, L.M. Boxt, and P.D. Murray, "Digital subtraction angiography: Overview of technical principles," Amer. J. Roentgenol. **139**, 781–786 (1982).

28. D.K. Swanson, P.D. Myerowitz, J.O. Hegge, and K.M. Watson, "Arterial blood-flow waveform measurement in intact animals: new digital radiographic technique," Radiology **161**, 323–328 (1986).

29. J.H. Bürsch, "Use of digitized functional angiography to evaluate arterial blood flow," Cardiovasc. Intervent. Radiol. **6**, 303–310 (1983).

30. K. Kedem, K. Kedem, C.W. Smith, R.H. Dean, and A.B. Brill, "Velocity distribution and blood flow measurements using videodensitometric methods," Invest. Radiol. **13**, 46–56 (1978).

31. N.R. Silverman and L. Rosen, "Arterial blood flow measurement: assessment of velocity estimation methods," Invest. Radiol. **12**, 319–324 (1977).

32. M. von Spreckelsen and K. Wolschendorf, "A method to determine the instantaneous velocity of pulsatile blood flow from rapid serial angiograms," IEEE Trans. Biomed. Engng. **BME-32**, 380–385 (1985).

33. J.N.H. Brunt, D.A.G. Wicks, D.J. Hawkes, A.M. Seifalian, G.H. du Boulay, A.F.C. Colchester, and A. Wallis, "The measurement of blood flow waveforms from x-ray angiography. Part 1: Principles of the method and preliminary validation," Proc. Instn. Mech. Eng. (Part H) **206** (H2), 73–85 (1992).

34. W.A. Bateman and R.A. Kruger, "Blood flow measurement using digital angiography and parametric imaging," Med. Phys. **11**, 153–157 (1984).

35. B.M.T. Lantz, J.M. Foerster, D.P. Link, and J.W. Holcroft, "Angiographic determination of cerebral blood flow," Acta. Radiol. Diagnosis. **21**, 147–153 (1980).

36. J.F. Lois, N.J. Mankovich, and A.S. Gomes, "Blood flow determinations using digital densitometry," Acta Radiologica. **28**, 635–641 (1987).

37. B. Lantz, "Relative flow measured by roentgen videodensitometry in hydrodynamic model," Acta Radiol. Diagnosis. **16**, 503–519 (1975).

38. G.G. Ferguson, "Turbulence in human intracranial saccular aneurysms," J. Neurosurg. **33**, 485–497 (1970).

39. C.W. Kerber and D. Liepsch, "Flow dynamics for radiologists. II. Practical considerations in the live human," Amer. J. Neuroradiol. **15**, 1076–1086 (1994).

40. G.L. Wolf, D.D. Shaw, and H.A. Baltaxe, "A proposed mechanism for transient increases in arterial pressure and flow during angiographic injections," Invest. Radiol. **13**, 195–199 (1978).

41. J.A. Abbott, M.J. Lipton, T. Hayashi, and F.C.S. Lee, "A quantitative method for determining angiographic jet energy forces and their dissipation: Theoretic and practical implications," Cathet. Cardiovasc. Diagn. **3**, 139–154 (1977).

42. D.E. Williamson, "Experimental determination of flow equation in catheters for cardiology," Am. J. Roentgenol. Rad. Ther. Nucl. Med. **94**, 704–709 (1965).

43. R.N. Cooley and L.B. Beentjes, "An inquiry into the physical factors governing

the flow of contrast substances through catheters," Am. J. Roentgenol. Rad. Ther. Nucl. Med. **89**, 308–314 (1963).

4.9 Problems

4.1. Given the stream function $\psi = -Ar^4(\sin^2\theta - \cos^2\theta)$, where A is a constant and r and θ are the two-dimensional polar coordiates, determine the corresponding velocities in the radial and polar directions.

4.2. Express Bernoulli's law, given as

$$P_1 - P_2 = \frac{\rho v_2^2}{2}\left[1 - \left(\frac{v_1}{v_2}\right)^2\right]$$

in terms of the cross-sectional area of the two points within the vessel.

4.3. (A) Bernoulli's principle can be applied to a syringe to describe the dynamics of an injection. Assuming that 1 is the position within the body of the syringe and 2 is the position within the throat of the syringe or region prior to entrance into the needle, derive an expression of the velocity of fluid exiting the syringe.
(B) What is the one factor neglected from the above derivation that could significantly influence the velocity of the injectate?
(C) If the throat of the syringe is 1/4 the diameter of the body of the syringe, how does the flow rate vary between the throat and body of the syringe?

4.4. If temperature were considered in fluid flow experiments, how would a decrease in temperature affect fluid flow?

4.5. Assuming flow of an incompressible fluid, if the velocity measured at one point within a vessel is 40 cm/s, what is the velocity at a second point that is 1/3 the original radius?

4.6. An electromagnetic flowmeter measures flow according to the relation $V = Bdv \times 10^{-8}$, where V is the voltage (V), B is the magnetic field strength (G), d is the internal diameter of the artery (cm), and v is the velocity of the flowing blood (cm/s). Express the potential difference V in terms of the flow rate Q.

4.7. For an artery with radius 0.25 cm, a magnetic field density of 500 g and a blood flow velocity of 150 mm/s, what is the expected measured voltage?

4.8. Identify a potential source of error in flow measurements using an electromagnetic flowmeter.

4.9. Given the influence of conductivity of the arterial wall, how would the voltage measured by the electromagnetic flowmeter be altered?

4.10. In terms of the magnetic field B and the electric current i, what is the direction of the magnetic force of the current?

4.11. What are the three criteria where the expression for the voltage from the electromagnetic flowmeter is valid?

4.12. What is the kinetic energy per unit volume of blood that has a speed of 0.5 m/s? (Note: $\rho_{\text{blood}} = 1000$ kg/m^3.)

4.13. Can the stream function and velocity potential function be interchanged. If so, why?

4.14. Derive an expression for the time needed for a fluid of velocity v to traverse a vessel of length L.

4.15. (A) Given the pressure and flow wave forms:

$$P(t) = \sum_{n=0} P_n e^{i(\omega nt - \phi_n)},$$

$$Q(t) = \sum_{n=0} Q_n e^{i(\omega nt - \theta_n)},$$

where n corresponds to the nth harmonic, what is the impedance?
(B) How are the angles ϕ and θ defined?

5
The Physics of Stroke

5.1 Introduction

This chapter will be devoted to a description of the biophysical and hemo-dynamic interactions involved in the origin, development, diagnosis, and treatment of stroke. The primary objectives of this chapter are to provide the reader with knowledge of (1) the causes and developmental mechanisms of the two major types of stroke: ischemic stroke and hemorrhagic stroke; (2) cerebral hemodynamics of stroke with particular attention to experi-mental and clinical measurements of regional cerebral blood flow and hemodynamics of a stenosis within the common carotid artery bifurca-tion; (3) theoretical and experimental models for the scientific and clinical investigation of stroke; (4) diagnostic techniques for the identification of sources and physiological consequences of stroke; and (5) current thera-peutic methods for the prevention of stroke and the minimizing of brain damage or ischemia as a result of stroke.

Optimal function of the human brain is critically dependent on the flow of blood circulating through the brain. In the normal state, cerebral blood flow originates from the internal carotid and vertebral arteries, flows through the arterial interchange at the base of the brain known as the circle of Willis, and permeates through the brain tissue via intricate networks of capillary vessels before returning to the heart through the venous system. If blood flow is obstructed or impaired at any point within this vascular route, por-tions of the brain are deprived of oxygen, initiating a cascade of mechanisms that results in brain tissue ischemia and eventually infarction. *Ischemia* refers to brain tissue deficient of oxygen that can still recover from potential brain damage, while *infarction* refers to permanent tissue death. The extent of ischemia is dependent primarily on the proximity of the obstruction and the collateral flow to that region. Any disease or neurological insult (brain injury) resulting in the severe restriction or complete cessation of blood flow to the brain is commonly referred to as *stroke*.

Stroke is a form of cerebrovascular disease that afflicts approximately

500 000 people each year and places a substantial burden on national health care, costing an estimated \$23.2 billion.[1] In addition, the mortality rate from stroke is 150 000 deaths per year. Stroke typically mainfests itself in afflicted patients as the following symptoms: (1) sudden weakness or numbness of the face, arm, or leg on one side; (2) sudden reduction or loss of vision, especially if it affects one eye only; (3) loss or impairment of speech or diminished understanding of speech; (4) sudden severe headaches with no apparent cause; (5) unexplained dizziness, unsteadiness, or sudden falls, particularly in combination with the aforementioned symptoms.[1] Within the past several years, a campaign entitled *brain attack* was initiated to generate public awareness on the symptoms of stroke.[2] This label is in direct reference to the term *heart attack*. The public has become educated about the symptoms of heart attack and knows that when initial symptoms such as chest pain or shortness of breath occurs, proper emergency personnel should be notified and immediate medical intervention sought. It was reasoned that because of the success of the term *heart attack* in increasing the public's knowledge of cardiovascular disease, the term *brain attack* would generate a similar increase in public awareness of cerebrovascular diseases. The onset of stroke symptoms can be as mild as slight dizziness and can easily be disregarded as a passing spell, when in reality a minor stroke has occurred, possibly leading to serious consequences. As with cardiovascular disease, time is of the essence with cerebrovascular disease, which requires immediate medical attention to minimize the severity of neurological deficits and improve upon the probability for a satisfactory patient outcome. It is hoped that when *brain attack* becomes a household term, the public will become educated about strokes and the devastating consequences can be reduced substantially.

Given the magnitude and corresponding implications of stroke on the quality of life and public health care, an intense search is underway for underlying causes and mechanisms, particularly in this decade, the Decade of the Brain (1990–2000). The purpose of this book, and of this chapter in particular, is to identify, describe, and illustrate these biophysical and hemodynamic mechanisms predisposing a person to stroke which often form the basis for novel methods of diagnosis and therapy. Two types of stroke that have been identified and recognized clinically correspond to their characteristic mechanisms of flow obstruction or neuronal damage: hemorrhagic stroke[3] and ischemic stroke.[4]

Hemorrhagic stroke. Hemorrhagic stroke occurs as a result of a ruptured vascular lesion at a point within the cerebrovasculature, typically due to an aneurysm or a weakened blood vessel within an arteriovenous malformation. In each of these cases, rupture into the brain causes hemorrhage, volumetric filling into the surrounding space, and resultant compression of the surrounding tissues. Hemorrhagic stroke occurs primarily in the young and middle-aged populations and constitutes approximately 20% of all stroke cases. The leading cause of hemorrhage in the young and middle-aged populations is vascular lesions such as arteriovenous malformations and aneurysms. Of hemorrhagic stroke in these populations, hemorrhage due to rupture of arteriovenous malformations occurs in 29% of all stroke cases and ruptures due to aneurysms occurs in 10% of all stroke cases. Aneurysms and arteriovenous malformations will be described in detail in Chaps. 6 and 7, respectively.

The leading cause of hemorrhage in the elderly is spontaneous intercerebral hemorrhage. In the elderly population (>65 yrs), the blood vessels of the brain tend to be brittle and less distensible due to the cumulative effect of atherosclerotic deposition. Atherosclerosis, coupled with hypertension, which has been shown to be present in 66% of the elderly population, results in the spontaneous rupture of these vessels and hemorrhagic stroke.

Ischemic stroke. The other type of stroke is ischemic stroke. In ischemic stroke, cessation of blood flow occurs as a result of luminal obstruction of the major arteries associated with the cerebral circulation. In the majority of cases of ischemic stroke, the arteries responsible for the transport of blood supply to the brain become obstructed or clogged through various mechanisms, reducing the blood flow and tissue perfusion. Of the two types of stroke, ischemic stroke is the most common, accounting for approximately 80% of stroke cases, and will be the primary focus of this chapter. Regardless of the type of stroke, the origin of stroke can be based, to some extent, on biophysical and hemodynamic interactions and subsequent mechanisms between blood flow and the blood vessel.

5.2   Biophysical and Hemodynamic Mechanisms of Stroke

As a result of stroke, regardless of the type of stroke, reduced blood flow to the brain may lead to neurological deficits and possibly death. Although the consequences are similar, the biophysical and hemodynamic mechanisms behind the obstruction of blood flow are different. Under the general heading of stroke and its two major types, i.e., ischemic and hemorrhagic, there exist three specific stroke subtypes: thrombosis, embolism, and hemorrhage. Mechanisms for the three subtypes of stroke are based on five distinct biophysical and hemodynamic processes of blood flow obstruction: atherosclerosis, embolus, thrombus, hemorrhage, and vasospasm which will now be discussed in detail below.

5.2.1   Atherosclerosis

Atherosclerosis, commonly referred to as *hardening of the arteries* is a pathological process in which lipid or fatty deposits from the flowing blood

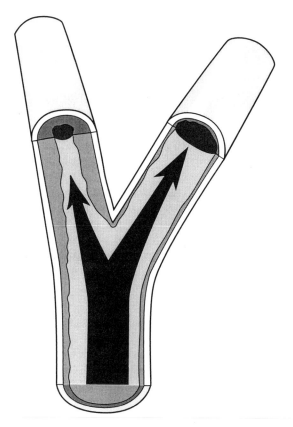

FIGURE 5.1. Schematic diagram of the distribution of atherosclerotic plaque along the inner wall of the carotid artery bifurcation.

accumulate along the innermost intimal layer of the vessel wall. Atherosclerosis and the development of arterial plaques are the product of a host of independent biochemical processes including the oxidation of low-density lipoproteins, formation of fatty streaks, and the proliferation of smooth muscle cells.[5] The presence of atherosclerotic lesions causes an irregular vessel surface and platelet aggregation due to turbulence, ultimately resulting in dislodgment into the bloodstream (Fig. 5.1). Atherosclerotic deposits promote the development of blood clots or the process of thrombosis due in part to flow obstruction and to high shear stresses exerted on the blood. Atherosclerotic thrombosis accounts for 33% of all stroke cases. Thrombosis will be explained later in this section. Further discussion of these processes is beyond the scope of this book. The interested reader should consult the following works by Ross,[6,7] Hajjar and Nicholson,[8] and Brown and Goldstein[9] for further information on biochemical processes and Gessner[10] and Fry[11] on hemodynamic and biophysical processes involved in atherosclerosis.

5.2.2   Embolus

An embolus represents gaseous or particulate (e.g., atheromata) matter that acts as traveling "clots." A common example of an embolus is a platelet aggregate dislodged from an atherosclerotic lesion. The dislodged platelet aggregate is transported by the bloodstream through the cerebrovasculature until it reaches vessels too small for further passage. The clot has no where to go and remains there, clogging the vessel and preventing blood flow from entering the distal vasculature. Although our discussion at the present is focused primarily on the carotid arteries and associated cerebrovasculature, emboli can originate from distant sources such as the heart, lungs, and peripheral circulation, which could eventually travel within the cerebral blood vessels, obstructing flow and causing stroke. Other sources of emboli include atrial fibrillation and valvular disease. The severity of stroke depends on the size of the embolus and the location of the obstruction. The bigger the embolus and the larger the vessel obstruction, the larger the territory of brain at risk. Approximately 31% of all stroke cases are attributed to emboli.

5.2.3   Thrombus

Thrombosis is an internal physiological mechanism responsible for the clotting of blood. In response to vessel or tissue injury, the blood coagulation system is activated, which initiates the following cascade of processes transforming prothrombin and resulting in a fibrin clot:

$$prothrombin \Rightarrow thrombin \Rightarrow fibrinogen \Rightarrow fibrin \Rightarrow fibrin\ clot.$$

Although there are a host of mechanisms and causes responsible for vessel injury, vessel injury can occur as a result of forces (shear stresses[11]) coupled with the excess energy created by the turbulent flow[12–14] exerted against the inner (intimal) lining of the vessel wall. Forms of therapy designed to treat patients with developed thrombi include antiplatelet therapy with ticlopidine and aspirin and "clotbusters" with a tissue-type plasminogen activator.[15]

5.2.4   Hemorrhage

With vessel rupture, hemorrhage occurs with blood seeping into the surrounding brain tissue or the subarachnoid space. As the blood accumulates within the brain, the displaced volume causes the blood, now thrombosed, to ultimately compress the surrounding vessels. The compression of vessels translates into a reduced vessel diameter and a corresponding reduction in flow in surrounding tissue, enlarging the insult. An additional effect of the displaced volume of hemorrhage is the increase of intracranial pressure, which could, in turn, translate to global neurological complications and dysfunction. Hemorrhage accounts for approximately 20% of all stroke cases.

5.2.5   Vasospasm

In the Introduction and the preceding section (Sec. 5.2.4), hemorrhagic stroke and resultant mechanisms were discussed. When bleeding occurs in the subarachnoid space, the arteries in the subarachnoid space can become spastic with a muscular contraction, known as cerebral vasospasm.[16] The contraction from vasospasm can produce a focal constriction of sufficient severity to cause total occlusion. The length of time that the vessel is contracted during vasospasm varies from hours to days. However, regardless of the duration of vessel constriction during vasospasm, reduction of blood flow induces cerebral ischemia, thought to be reversible within the first six hours and irreversible thereafter. It has been shown that vasospasm is maximal between five and ten days after subarachnoid hemorrhage and can occur up to two weeks following subarachnoid hemorrhage. The resultant damage to brain tissue can be minimized with the administration of pharmacological agents such as the vasodilator papaverine.

The causes and mechanisms of vasospasm have not been identified and understood clearly. Endothelin, a physiological agent that is the most potent and long-lasting vasoconstrictor ever isolated from mammalian cells, is among the agents believed to be responsible for cerebral vasospasm after subarachnoid hemorrhage.[17] Thus, strategies devised to combat vasospasm involve anatagonists or agents that work to counteract the effects of endothelin and vasoconstriction by inducing vasodilation in the afflicted vessel.[18,19] Determination of the clinical efficacy of these agents in the prevention of vasospasm and its resultant ischemic neurological deficits requires additional clinical testing in larger patient populations before widespread acceptance as an effective treatment.[20]

Since the majority of strokes that occur are ischemic in nature, the discussion presented below will address primarily the biophysical mechanisms involved in the development, diagnosis, and therapy of ischemic stroke. The extent of cerebral ischemia and infarction as a result of any of these sources of flow obstruction is due in part to the proximity of vascular occlusion and the collateral flow in the brain. A stenosis of the distal internal carotid artery or proximal middle cerebral artery will be more likely to cause neurological complications as opposed to a stenosis much more distally located within the brain. Since the carotid artery represents the entry level of cerebral blood flow, more of the brain will potentially be affected by a stenotic occlusion.

5.3   Hemodynamics of the Carotid Artery Bifurcation

As mentioned in Chap. 3, the common carotid arteries, which bifurcate ultimately into a vast and complex arrangement of cranio-facial-cerebral blood vessels, represent an important vascular location for atherogenesis.

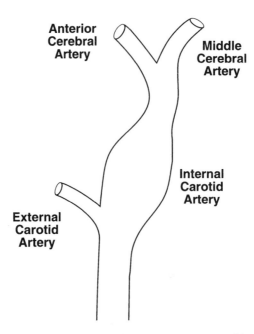

FIGURE 5.2. Schematic diagram of the human common carotid artery bifurcation, illustrating the curvature and bifurcation geometric features making it susceptible to cerebrovascular disease.

One reason for this is the geometry presented by the carotid artery bifurcation. The carotid artery bifurcation, typically represented as a Y-shaped structure is, in reality, a much more complicated structure (Fig. 5.2). The two geometric features of the carotid artery that make it more hemodynamically vulnerable to potential processes causing stroke are vessel curvature and the arterial bifurcation apex.

5.3.1   Hemodynamics at the Vessel Curvature

There exist a number of points along the human carotid artery that exhibit curvature, particularly before and following the bifurcation apex. The importance of vessel curvature in the carotid artery can be demonstrated by considering the hemodynamic parameter wall shear stress. As mentioned in Chap. 4, the tails of the velocity profile (the points of the profile in contact with the vessel wall) exhibit the maximum wall shear stress. As blood flows through a curved portion of a blood vessel, the velocity profile becomes skewed away from the radius of vessel curvature. This phenomenon occurs in common everyday experiences such as driving a car or bicycle around a curve or sledding in a luge (Fig. 5.3). Either example typically involves circular motion of an object or person that is maintained by a centripetal or

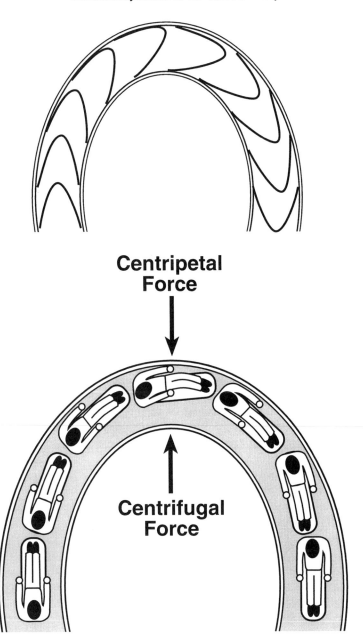

FIGURE 5.3. Schematic diagram illustrating the physical interactions experienced by an object in angular or circular motion. (A) The velocity profile of blood flow about a curved vessel becomes skewed at vascular points of curvature. (B) The reasons behind the skewed velocity profile can be likened to the centrifugal force, acting outward in a radial direction. The centrifugal force is counteracted by a centripetal or "center-seeking" force.

"center-seeking" force given by

$$F_{\text{centripetal}} = ma = \frac{mv^2}{R}.$$

As R decreases (e.g., the curve becomes sharper), the centripetal force increases accordingly. The centripetal force is counteracted by a centrifugal force of the same magnitude that is always directed outward from and perpendicular to the axis of rotation. When driving around a curve, one leans into the curve to counteract the centrifugal force. It is the centrifugal force acting against the blood particles that causes a shift in the center of mass and hence the velocity profile. Once the blood flow passes the curve, the influence of the centrifugal force is removed and the velocity profile returns to its axisymmetric form.

5.3.2   Hemodynamics at an Arterial Bifurcation Apex

An arterial bifurcation, the normal geometry of the common carotid artery termination, consists of a main or parent artery branching into two daughter arteries. Blood flows from the parent artery, is separated at the bifurcation apex or branching point, and is diverted into the daughter arteries. The apex is subjected to the impact of the flowing blood, its magnitude, and the resultant effects that are dependent on its kinetic energy. If the flow energy level is high enough, vessel wall damage and subsequent degeneration may occur.[21] As the blood flow enters the daughter arteries, the velocity profile is dramatically skewed toward the apical region of the bifurcation, implying a significant increase in wall shear stress at the vessel wall. The carotid artery, although sufficiently approximated by a Y-shaped geometry, possesses a localized distension or pocket of the vessel, known as the carotid sinus or bulb. From a hemodynamic standpoint, the carotid sinus serves as a source of recirculation (vortex flow), the formation and size of which depends on the Reynolds number and the ratio of flow between the parent and daughter artery.[22] The hemodynamics at the carotid artery bifurcation is depicted in the angiogram and ultrasonic image in Fig. 5.4.

In addition to the kinetic energy imparted to the bifurcation apex, there is accompanying turbulence. The arterial bifurcation affects the hemodynamic properties of the flowing blood by lowering the Reynolds number in some cases by half depending on the bifurcation angle or the angle between the two daughter arteries.[23] The reduction of the Reynolds number, as described in Chap. 4, increases the likelihood of turbulence. A final implicating factor of the bifurcation to the propensity of stroke is the correlation of the geometry and hemodynamics of the carotid artery bifurcation to the onset and development of atherosclerosis.[24,25] The premise of such a hypothesis involves the highly variable regions of wall shear stress characteristic of flow separation. Regions of elevated shear stress are believed to cause injury to the endothelial cell layer of the vessel wall and predispose the

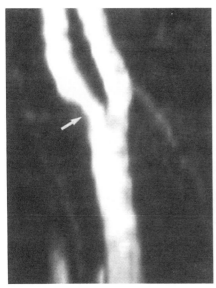

FIGURE 5.4. MR angiogram and color Doppler image (see color insert that follows p. 170) depicting anatomical and hemodynamic features of a normal carotid bifurcation. (Reprinted, with permission, from *Radiology* **185**, R. L. Wolf, D. B. Richardson, C. C. LaPlante, J. Huston, III, S. J. Riederer, and R. L. Ehman, Blood flow imaging through detection of temporal variations in magnetization, pp. 559–567. © 1992 by the Radiological Society of North America.)

vessel to disease. The body responds to the injury with platelet aggregation and clotting mechanisms, creating the potential for stroke.

5.4   Hemodynamics of a Stenosis

The preceding section discussed anatomic and hemodynamic factors presented by the carotid artery bifurcation in its normal state that could predispose an individual to atherosclerosis. The implementation of vascular disease such as atherosclerosis further complicates and compounds the potential dangers implicated in stroke. The carotid artery and hemodynamic effects were considered on a global issue; however, we would now like to focus our discussion on the local hemodynamic effects presented by an occlusion or stenosis.

The luminal reduction or occlusion presented by a vascular stenosis introduces significant changes in the vessel geometry as well as the hemodynamics before, during, and following the formation of stenosis. The geometry of a stenosis presents an irregularity in the vessel contour, the consequences of which manifest themselves into alterations of the normal hemodynamics.

In addition to the size and general shape of the stenosis, the factor of most interest is the maximum reduction in vessel diameter caused by the stenosis. Assuming that D_{norm} is the diameter of a normal region of the vessel and D_{sten} is the minimum residual diameter of the vessel caused by the stenotic lesion, clinical assessment of carotid artery disease is based on the percent stenosis (% stenosis) presented by a lesion:

$$(\% \text{ stenosis}) = \frac{D_{norm} - D_{sten}}{D_{norm}} \times 100\% = \left(1 - \frac{D_{sten}}{D_{norm}}\right) \times 100\%$$

with the following categories: mild (1%–39%), moderate (40%–59%), severe (60%–79%), critical (80%–99%), and occluded (100%).[26] This point represents the foundation for all physical and clinical measurements relating to the severity of the stenosis and indications for treatment options.

Hemodynamics through an arterial stenosis is a problem of paramount importance in discussions of cerebrovascular disease and has been studied extensively.[27–32] The basis for the hemodynamics at a vascular stenosis is Poiseuille's law, which, in effect, states a linear relationship between volumetric flow rate and pressure gradient under normal circumstances, and Bernoulli's principle, which states an inverse relationship between pressure and flow velocity. Let us first consider Poiseuille's law of fluid flow. The linear relationship holds true only as a first-order effect and is valid only to a "point" where the hemodynamic relationship becomes nonlinear or drastically changes in form. The validity of Poiseuille's law of fluid flow is compromised when the fluid becomes non-Newtonian or turbulent, or when the vessel radius is restricted to the stenotic region as opposed to the entire vessel.[33] One is then left to consider quantitative values of flow rate as the pressure gradient increases continually within the stenosis past the point of nonlinearity. More specifically, the reasons behind the inapplicability of Poiseuille's law include: (1) non-Newtonian viscosity of blood, (2) turbulence, (3) pulsatile driving pressures, (4) kinetic energy transformations, and (5) distensibility of vessels.[34] This "point" corresponds to the transition from laminar flow to turbulent flow. Byar et al.[33] performed a series of fluid flow experiments investigating the influence of all hemodynamic parameters contained within Poiseuille's law and found that fluid flow Q, as it pertains to a stenosed vessel, is related to the pressure gradient ΔP, according to

$$Q = \left(\frac{\Delta P}{10^a}\right)^{1/b},$$

where b is a constant relating the change of ΔP with respect to Q (the slope of the experimental flow curve) and a is the value of ΔP at zero Q (graphical intercept of the experimental flow curve). Interestingly enough, although they were carefully considered in experimentation, there is no external dependence on fluid viscosity, vessel radius, and vessel length. The influence

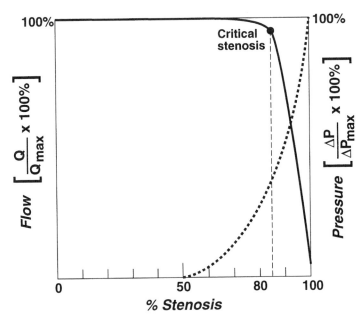

FIGURE 5.5. Graph showing the changes of pressure gradient and flow through progressive luminal obstruction of a vessel. The critical stenosis denotes the point at which a marked reduction of flow and a corresponding increase in pressure gradient are observed.

of these factors, however, are incorporated into the mathematical function and associated graphical constants a and b.

As the stenotic region of a vessel becomes critical, the flow rate decreases, the pressure gradient across the stenosis decreases, and the flow velocity increases as illustrated by Bernoulli's principle. In addition, there is a buildup of excess pressure proximal to the stenosis. The trends of these hemodynamic parameters continue until a critical stenosis is reached. The critical stenosis is defined as the percent stenosis at which intravascular flow approaches zero and the prestenotic pressure approaches its maximum value. The critical stenosis is unique to vessel geometry and hemodynamics but has been shown to occur generally at about 80%–85% obstruction of the major vessels in the human vasculature.[35–38] At the point of critical stenosis, a sharp decrease of flow rate is observed due to the increased turbulence proximal to the stenosis, as shown in Fig. 5.5.

As the stenosis progresses to occlusion, the pressure drop across the stenosis reaches 100% of the maximum and the flow rate is zero. In addition, the prestenotic pressure is equal in magnitude to that at the origin of the parent vessel. As an illustrative example of Newton's third law, the stenosis exerts an equal yet opposite force against the hemodynamic forces generated by the systemic blood pressure driving blood through the obstructed vessel. At

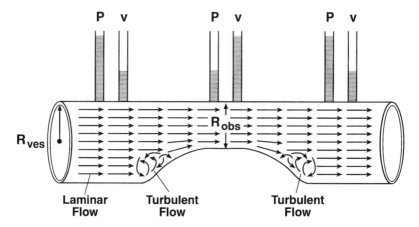

FIGURE 5.6. Schematic diagram illustrating Bernoulli's principle and the changes of pressure and velocity in a vessel containing a stenosis.

100% stenosis, the pressure gradient is not exactly zero due to the additional energy losses. However, no relations, experiments, or adequate explanations exist that elucidate and accurately characterize these energy losses leading one to approximate the pressure gradient.

Bernoulli's principle, first introduced in Chap. 4, describes the total energy of a flowing fluid through a rigid vessel:

$$E = P + \frac{\rho v^2}{2}.$$

According to Bernoulli's principle, the flow velocity v per unit volume of fluid is inversely proportional to the intravascular pressure P, which can be determined by Poiseuille's law. The premise behind applications of Bernoulli's principle to a stenosed vessel is that, since conservation of mass holds, Bernoulli's principle can be applied to two points representing flow through different segments of a vessel. Therefore, Bernoulli's principle can be expressed for the two points according to the following: one point (1) representing flow through a normal unobstructed region of the vessel and another point (2) representing flow through an obstructed region at the maximum point of stenosis. Bernoulli's principle expressed for these two points can be equated, relating hemodynamic parameters characteristic of the two distinct regions of flow:

$$P_1 + \frac{\rho_1 v_1^2}{2} = P_2 + \frac{\rho_2 v_2^2}{2}.$$

The underlying physical interactions of Bernoulli's principle are illustrated in Fig. 5.6. From a biophysical standpoint, we now have an expression for the loss of pressure due to a partial obstruction of the vessel. The ability to make such calculations is important due to the probability that clinical measurements and observations at or around a stenosis are susceptible to errors due to the small distances imposed by stenoses and limitations in the imaging modality and corresponding technique.

Returning to our initial discussion on Bernoulli's principle, the inclusion of viscous forces into the fluid flow problem requires modifications to the original equation describing Bernoulli's principle. Incorporating terms representing the contribution of stenosis length to pressure drop across stenosis and the contribution of distal luminal expansion of the pressure drop, the equation of a pressure drop across the stenosis is[39]

$$\Delta P = \frac{8\mu}{R^2} L v_1 \left(\frac{A_1}{A_2}\right)^2 + \rho v_1^2 \left(\frac{A_1}{A_2}\right)^2,$$

where μ is the blood viscosity, L is the vessel length, R is the vessel radius, A_1 is the cross-sectional area of a normal arterial lumen, A_2 is the cross-sectional area of a stenosed arterial lumen, v_1 is the velocity of blood in an unstenosed artery, and ρ is the density of blood.

Since ischemic strokes are the most common and most likely the result of atherosclerotic lesions developed along the arterial wall, we will now consider the influence of an atherosclerotic lesion on the vessel wall and corresponding interactions between the flowing blood and vessel wall. As mentioned earlier, an atherosclerotic lesion is an irregularly distributed mass of calcified fatty deposits that narrow the arterial lumen and stiffen the affected portion of the vessel wall, creating a region of rigid tissue countered on either end by vascular wall that has retained its elastic behavior. This causes significant changes in the biophysical and biomechanical properties of the vessel, resulting in reduced distensibility.

5.4.1 Reduced Volumetric Blood Flow Rate

The first and probably most obvious of these biophysical changes caused by flow through the stenotic lesion is the reduction of blood flow. The consequences of reduced blood flow are a possible decrease in blood flow volume circulating through the brain and the occurrence of stasis and thrombosis. As the blood flow capacity of the affected artery is reduced, the other carotid artery accommodates by increasing blood flow and maintaining adequate levels of brain tissue perfusion. However, problems originate at the site of the occlusion. Blood is a fluid that must be in continual motion in order to function properly. As blood flow is reduced at the stenosis, the flow becomes stagnant and clotting mechanisms are invoked, and the blood begins to clot within the vessel. The mechanism by which blood clots as a result of reduced motion is termed stasis and the resultant clot is the thrombus. It should be noted that the thrombus does not adhere strongly to the vessel wall and itself can be dislodged into the bloodstream as an embolus and result in stroke.

5.4.2 Increased Blood Flow Velocity

The effects of increased blood flow velocity occur according to three different mechanisms:

1. Increased blood flow velocity induces a large kinetic energy at the stenosis, exerting a significant hemodynamic force against the normal portion or post-stenotic region of the vessel wall. The increased blood flow velocity through the stenosed region of the abnormal vessel exhibits unique characteristics and is termed *jet flow*. Jet flow, commonly used to describe flow exiting a hypodermic needle or catheter under fairly large pressures (see Chap. 4), represents the turbulent nature of flow following a constricted area of the vessel. Prolonged impingement of the blood flow at this magnitude of force could induce structural fatigue and corresponding changes in the vessel wall, resulting in distension and ultimately leading to the development of an aneurysmal dilatation (Chap. 6). The distension distal to the stenosis, also known as post-stenotic dilatation, is believed to be due, in part, to the conversion from a large kinetic energy to a large potential energy, as given by Bernoulli's principle.

2. Increased blood flow velocity coupled with the irregular geometry causes a decrease in Reynolds number, and a corresponding tendency for the blood flow to become turbulent. In turbulent flow, the kinetic energy produced by the flowing blood is transferred into the cracks and crevices presented by the abnormal plaque distribution with potentially enough accumulated energy over time to dislodge a portion of the plaque into the bloodstream where it now becomes a particulate embolus (Fig. 5.7). In addition, the developed turbulence is believed to be the source of *bruits*, or audible sounds, detected with a standard stethoscope placed over the stenosed area.[40,41] Although the bruit frequencies vary, a recent study by Kurokawa et al.[42] revealed a frequency less than 850 Hz for arterial stenoses less than 70% and greater than 800 Hz when the stenosis was more than 70%.

3. Increased blood flow velocity causes a high shear stress along the upper portion of the lesion and a region of low shear stress along the tails and bottom portion of the lesion. The shear stress acts in conjunction with the kinetic energy created by the turbulent flow to create a potentially dangerous situation. Another hemodynamic phenomenon that occurs as the direct result of increased blood flow velocity due to a sudden change in the vessel

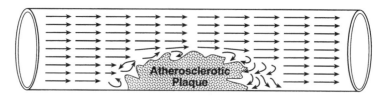

FIGURE 5.7. Illustrative example of Bernoulli's principle, implicating hemodynamics as a possible mechanism for plaque dislodgement and ischemic stroke.

diameter is the water hammer effect. In reality, it is not the change in blood flow velocity but the change in pressure that forms the basis for the water hammer effect. As the fluid impinges on the constricted area of the vessel, the hemodynamic energy is expended in forcing the fluid through the constriction and distensions of the vessel wall, causing the rapid changes in pressure, resulting in the water hammer effect. The sudden constriction also causes a reflection of the hemodynamic pressure wave, propagating in the opposite direction with a wave velocity c given by[43]

$$c = \sqrt{\frac{Eh}{2\rho R}},$$

where E is the Young's modulus of elasticity, h is the wall thickness; R is the mean vessel radius, and ρ is the blood density. This is known as the Moens–Korteweg equation. As the elastic modulus of the atherosclerotic vessel decreases, so does the wave velocity, resulting in a localized deposition of kinetic energy proximal to the lesion.

5.4.3   Compliance Mismatch

The changes in elasticity induced by the atherosclerotic lesion is termed *compliance mismatch*. Compliance, an indirect measurement of the vessel wall elasticity, is the change in volume with respect to the change in pressure or $C = \Delta V / \Delta P$. The implications of a compliance mismatch can be visualized by considering the elastic function in response to blood flow. As the pulsatile flow strikes the wall of a normal vessel, the elastic wall reacts with a recoil in response to the hemodynamic forces, further propelling the blood along the vasculature. In the atherosclerotic region of the vessel, the recoil response is substantially reduced or eliminated, depending on the extent and distribution of the lesion, which may not be sufficient to force blood around the lesion. This also translates in the body having to exert more force and produce more work in order to maintain proper levels and rates of blood flow.

As the incoming hemodynamic pressure wave propagates through the blood vessel, containing the atherosclerotic lesion, part of the wave is transmitted through the patent portion of the vessel, while the remaining part of the wave reflects off the lesions and propagates in a direction opposite to that of the original incoming pressure wave. The reflected pressure wave is critically damped within the boundaries created by the normal and atherosclerotic lesion of the vessel wall. The increase in damping reduces the natural frequency of the vessel wall. Since the natural frequency of the normal vessel is approximately 1–2 kHz and the frequency of the pulsatile blood flow is about 450 Hz, the vessel wall frequency could be reduced, depending on the distribution and extent of the stenotic lesion, to equal that of the blood flow, making resonance and subsequent rupture a physical possibility. Setting aside issues of geometry for the moment, the vibrational displacement of an elastic object subjected to a periodic driving force can be expressed mathematically by the differential equation

$$A \frac{d^2 x}{dt^2} + B \frac{dx}{dt} + Cx = P_0 \cos \omega t.$$

This is the same differential equation seen in Chap. 3 to describe the physical basis for the manometer with the elastic membrane. The solution of the above equation, the step-by-step derivation that can be found in any elementary engineering textbook,[44] yields the following expression for the resonant frequency ω_r:

$$\omega_r = \sqrt{\frac{C}{A}\left(1 - \frac{B^2}{2AC}\right)}.$$

The consequences of the resonant frequency to the problem at hand can be seen by increasing the damping coefficient B. Increasing B lowers the resonant frequency of the atherosclerotic wall, making resonance more likely.

5.5   Theoretical and Experimental Models of Stroke

As with all good scientific investigations, a study of the biophysical and hemodynamic mechanisms of stroke, in general, requires the use of theoretical and experimental models to advance current knowledge on the etiology of the disease and foster the development of novel approaches of diagnosis and therapy in stroke. Theoretical models of stroke involve the computational analysis of fluid dynamics through a vessel with a constriction implemented into the fluid flow geometry. In a typical computational simulation, the stenosis can be implemented by the following mathematical form:

$$y = A \exp\left(-\frac{(x-B)^2}{A^2}\right),$$

where A is the maximum extension of the stenosis into the vessel lumen and B is the location of the stenosis peak along the longitudinal axis of the vessel (Fig. 5.8). Computational analysis of vascular hemodynamics in stroke has been performed successfully through investigations of steady flow[45–47] and pulsatile flow[48,49] through a vessel with a vascular stenosis. Similar *in vitro* experiments have been performed, correlating well with the results from the computational simulations.[47,50–52]

Experimental *in vivo* animal models of stroke have been created successfully in rats, gerbils, cats, and baboons.[53] Atherosclerotic lesions are produced by feeding the animal an atherogenic diet consisting of a diet rich in lipids. This diet over an extended period tends to promote the development of fibrous or fatty plaque, resulting in an atherosclerotic stenosis.

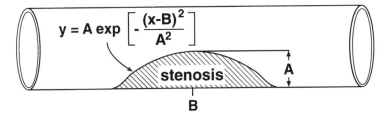

FIGURE 5.8. Schematic diagram depicting the dimensions of a stenosis typically implemented in theoretical and experimental studies.

5.6 Diagnostic Techniques for Stroke

A number of medical imaging modalitites have been applied successfully in the acquisition of visual and quantitative information on the structure and hemodynamic or metabolic function of the brain, particularly as a result of ischemic cerebrovascular disease.[54-56] These imaging modalities and corresponding techniques as applied to the brain or head and neck region fall within the field of radiology known as neuroradiology. Neuroradiological imaging procedures are utilized basically in two distinct capacities. The first capacity is the diagnostic assessment of the brain for the presence or absence of cerebrovascular disease. A patient complaint of neurological symptoms such as headaches, double or blurred vision, fainting spells, dizziness, and seizures may be followed up with a neuroradiological imaging procedure regardless of whether a direct cause is apparent or evident. The choice of imaging procedure is dependent generally on the nature and severity of the symptoms and is also based on the experience of the attending physician.

Neuroradiological diagnosis serves a purpose not only in the determination and identification of a cerebrovascular disease but also in the assessment of therapeutic options and strategies for the treatment of such diseases. Thus in an attempt to address these purposes, it should also be mentioned that there are, in effect, two types of diganostic assessment available from neuroradiological imaging procedures: anatomical and physiological. Anatomical assessment of cerebrovascular diseases allows one to identify a vascular stenosis or hemorrhage and the involved vascular region and to quantitate the size of the cerebrovascular lesion such as an aneurysm (Chap. 6) or arteriovenous malformation (Chap. 7). Figure 5.9 shows an angiogram of a carotid artery stenosis. Physiological assessment allows one to determine the severity and regions of reduced cerebral blood flow in the brain as a result of occlusive carotid artery disease, to determine the presence and degree of collateral flow, and to investigate the metabolism and viability of afflicted brain tissue.

The second capacity involves the use of selected neuroradiological imaging procedures as a means of guiding interventional or endovascular therapy. These selected imaging procedures, e.g., digital subtraction angiography,

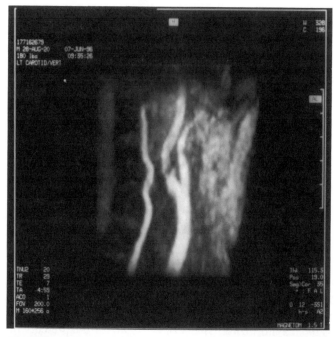

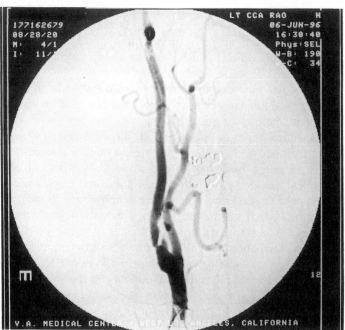

FIGURE 5.9. Angiograms of a carotid artery stenosis.

permit the continual visualization and evaluation of the cerebrovascular region containing the disease with the use of catheters navigated through the vasculature and injection of contrast medium to assess circulatory physiology (blood flow) within the region. In a typical diagnostic or therapeutic procedure, catheters (long, thin, flexible tubes) are introduced into a major artery of the human body, generally the femoral artery located in the upper leg, and guided through the arteries of the body to the region of the cerebrovascular disease, e.g., vessel stenosis or occlusion, aneurysm, or arteriovenous malformation, as shown in Fig. 5.10. Injection of contrast medium at a point proximal to the cerebrovascular lesion allows one to visualize critical information concerning the structure and, to a lesser extent, function of the lesion.

With regard to the clinical consequences of stroke, a host of medical imaging modalities have been employed successfully in the diagnostic assessment of cerebral hemodynamics and the presence and quantitative extent of regions of cerebral ischemia. These modalitites include computed tomography[57,58] (CT), magnetic resonance imaging[59-62] (MRI), magnetic resonance angiography[63] (MRA), magnetic resonance spectroscopy[64-67] (MRS), ultrasound imaging[68,69] (US), digital subtraction angiography[69,70] (DSA), and nuclear medicine techniques [single-photon emission computed tomography[71-75] (SPECT) and positron emission tomography[76] (PET)]. The importance of imaging techniques in the diagnostic assessment of stroke is illustrated in Fig. 5.11. For detailed information on the operating principles and general theory behind each of these imaging modalities, the reader is referred to the cited references for additional reading. We would, however, like to elaborate on the physics behind ultrasound imaging because of its far-reaching applications in obtaining both anatomical and physiological information critical in the clinical assessment of stroke.

5.6.1 The Physics of Ultrasound Imaging

In contrast to the other imaging modalities presented in this chapter, ultrasound can be used to obtain visual anatomic images of the carotid bifurcation as well as the vessels within the brain in addition to images of anatomical structures, such as the carotid arteries, that are physiologically relevant to brain function. When applied to the carotid arteries in the neck, ultrasound reveals quantitative information concerning size of a vessel stenosis as a result of atherosclerotic deposits and blood flow through the stenosis. Patients presenting with a stenosis corresponding to a $\geq 70\%$ diameter reduction are candidates for surgery regardless of the presence of symptoms. Unless diagnosed and treated in the early stages, stenosed vessels indicate a high risk for stroke.

Ultrasound imaging utilizes sound waves in the ultrasonic range (20 Hz–20 kHz) to obtain both anatomical and physiological information. In an ultrasound imaging procedure, an ultrasonic transducer is placed directly over the region of interest. The transducer generates sounds waves that

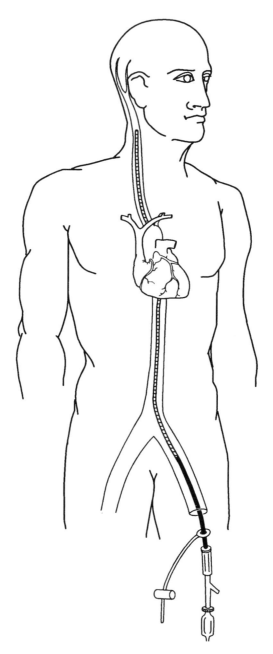

FIGURE 5.10. Schematic diagram illustrating the fundamental technique of diagnostic and therapeutic embolotherapy.

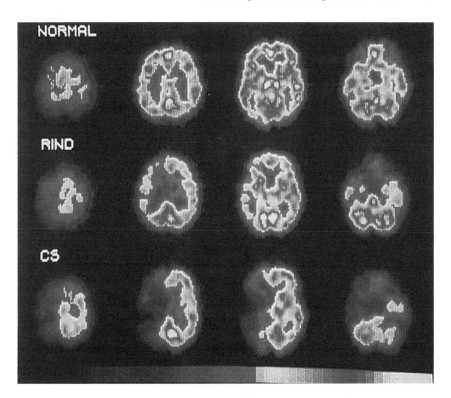

FIGURE 5.11. Single-photon emission computed tomographic images (see also color insert that follows p. 170) that represent various stages of cerebral ischemia as indicated by the reduction in cerebral blood flow. (Reproduced, with permission, from *Stroke*, **20**, I. Podreka, C. Baumgartner, E. Suess, C. Müller, T. Brücke, W. Lang, F. Holzner, M. Steiner, and L. Deecke, Quantification of regional cerebral blood flow with IMP-SPECT: reproducibility and clinical relevance of flow values, pp. 183–191. © 1989 American Heart Association.)

propagate from the transducer through the surface of the patient. Depending on the physical nature of the wave, the sound interacts with the various tissues and organs in one of two possible ways: reflection off the tissue–organ interface back toward the transducer for detection or transmission through the tissue medium. The mechanism of ultrasonic interaction with tissue is dependent primarily on the characteristic impedance of the medium or the particular tissue or organ and is expressed in terms of coefficients or ratio of the reflected, A_r, and transmitted, A_t, amplitudes of the pressure wave with respect to the amplitude of the incident wave, A_i:

$$\frac{A_r}{A_i} = \frac{Z_2 \cos \phi_i - Z_1 \cos \phi_t}{Z_2 \cos \phi_i + Z_1 \cos \phi_t},$$

$$\frac{A_t}{A_i} = \frac{2Z_2 \cos \phi_i}{Z_2 \cos \phi_i + Z_1 \cos \phi_t},$$

TABLE 5.1. Ranges of peak systolic velocity (PSV) observed at the internal carotid artery for corresponding ranges of stenosis.

% Stenosis	PSV (cm/s)
0–50	< 125
50–75	125–250
75–90	> 250
> 90	Doppler less reliable as stenosis approaches occlusion

where Z_1 and Z_2 represent the characteristic impedances of the two media and ϕ_i and ϕ_t, are the angles of the incident and transmitted wave, respectively, with respect to an axis perpendicular to the interface. The reflected sound propagates back toward the transducer in the form of "echoes" and are subsequently detected, registered, and converted to electrical signals. The electrical signals represent the dynamic images captured by the transducer and are visualized on a video display.

Physiological information such as blood flow velocity is measured by ultrasound incorporating the principles of the Doppler effect. Ultrasonic waves of an original frequency v_0 and corresponding speed V are focused in a continuous beam on the surface of a blood vessel, interact with the red blood cells contained within the flowing blood, and are reflected back toward the transducer at an altered, Doppler-shifted frequency v_D with a velocity v_D. The Doppler-shifted frequency is related to the aforementioned parameters by the relation

$$v_D = \frac{2v_0 V}{v_D} \cos \theta,$$

where θ is the angle between the axis of sound propagation and the flowing blood.

In clinical practice, a carotid color flow duplex study (ultrasound and Doppler assessment) is the first diagnostic study ordered when a patient presents with either a minor stroke commonly referred to as transient ischemic attack (TIA) or a completed stroke. Fifty percent of strokes are caused by atherosclerotic disease around the common carotid bifurcation.[77] At duplex examination, the common carotid artery bifurcations are evaluated for atherosclerotic plaque using gray-scale images, and the degree of stenosis is determined indirectly from the Doppler analysis (Table 5.1). The ideal angle of insonation is 0°, where $\cos \theta = 1$. The angle should never exceed 60° because of inherent inaccuracies. Studies have shown that when a diameter stenosis $\geq 70\%$ is present, there is a significantly increased risk of stroke as a result of decreased blood flow to the brain.[78] Therefore, the goal of the duplex examination is to identify those patients with diameter stenoses in the 75%–90% range as surgical candidates for carotid endarterectomy, regardless of the presence, nature, or severity of symptoms.

Transcranial Doppler is a noninvasive method by which vessels around the circle of Willis can be imaged and used to detect intracranial carotid stenosis[79] and vasospasm resulting from subarachnoid hemorrhage.[80] It can assess physiological changes in blood flow as a result of various disease states in which peak systolic velocities (PSV) and resistive indices are increased and decreased, respectively.[81] Because of a lack of sensitivity and specificity, this noninvasive test is best used in conjunction with cross-sectional imaging and digital subtraction angiography.

5.7 Assessment of Risk for Stroke

Risk assessment of a patient for stroke is based typically on a set of risk factors, categorized as nonmodifiable and modifiable. Risk factors assist as a subjective means of identifying individuals who are at greater risk for stroke than others. The term *modifiable* refers to a person's ability to effectively address the particular risk factor and reverse the probability of the occurrence of stroke. For example, nonmodifiable factors for stroke include age, gender, race, and heredity. Each of these factors have been shown, in given studies, to increase the risk for stroke to varying degrees and are unique to a specific individual. Nonmodifiable risk factors cannot be changed or modified in any instance. There exists another set of risk factors, modifiable risk factors, which have shown to increase a person's risk for stroke but can be modified by medication, diet, exercise, or lifestyle modification. Modifiable risk factors include hypertension (modified by antihypertensive medication and/or diet), cardiac disease (modified by medical or surgical intervention, medication, diet, or exercise), cigarette smoking (modified by cessation), and massive alcohol consumption (modified by drastic reduction or cessation). Other factors such as diabetes melittus, previous stroke, and the presence of other diseases are suspected as relevant risk factors for stroke and are currently under investigation. These factors form the basis or framework for the development of a tailored profile concerning a patient's inherent risk for stroke.

Although a certain patient may conform to every single category listed above and be a prime candidate for stroke, most patients will not seek medical attention until pathophysiological changes become symptomatic, prompting the patient to seek medical attention. Upon diagnosis, the physician assesses the severity of the clinical condition and the subsequent risk or potential for manifestation to stroke. Thus, risk assessment of stroke also depends on the clinical presentation of the cerebrovascular lesion at diagnosis. In the setting of intracranial hemorrhage, aneurysms and arteriovenous malformations are sought. As previously stated, these lesions will be discussed further in Chaps. 6 and 7, respectively. It should be stated, however, that because these lesions are situated within the brain, developmental growth may in itself elicit neurological symptoms, prompting medical attention and subsequent diagnosis. In other words, these lesions may be symptomatic prior to hemorrhage, allowing the attending physician to properly

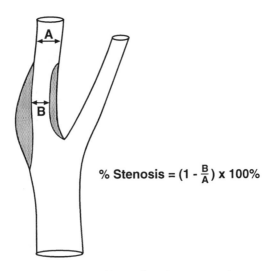

$$\% \text{ Stenosis} = (1 - \frac{B}{A}) \times 100\%$$

FIGURE 5.12. Schematic diagram illustrating the commonly accepted method of measuring carotid artery stenosis. (Reproduced, with permission, from *Journal of Vascular Surgery* **8**, J. D. Baker, R. B. Rutherford, E. F. Bernstein, R. Courbier, C. B. Ernst, R. F. Kempczinski, T. S. Riles, and C. K. Zarins, Suggested standards for reports dealing with cerebrovascular disease, pp. 721–729. © 1988 Mosby-Year Book, Inc.)

assess risk of hemorrhage in an attempt to avert rupture and its catastrophic consequences.

For ischemic (embolic or atherosclerotic) stroke, on the other hand, the primary point of concern may be the presence and extent of extracranial carotid artery stenosis. Although a variety of methods exist for obtaining accurate two- and three-dimensional images of the carotid artery and associated stenosis, there still exists considerable controversy regarding the quantitative measurement of carotid artery stenosis.[82–86] A generally accepted method for the measurement of carotid artery stenosis is displayed schematically in Fig. 5.12. At issue is the resolution limitations of medical imaging modalities. Since the suspected vascular lesion is not located within the brain, it becomes less likely that symptoms will present as a result of the stenosis, making diagnosis of the lesion prior to episodes of stroke difficult. Additional factors such as family or medical history, particularly in areas of cardiovascular health, tend to be strong indicators as to the likelihood of carotid artery disease and stroke.

5.8   Treatments for Stroke

Therapeutic options are considered and staged according to two different scenarios: (1) prevention of stroke and (2) minimizing neurological effects of cerebral ischemia following a stroke.[87] For the prevention of stroke, upon identification of an atherosclerotic lesion along the carotid artery, the primary objective is to remove the risks of (a) further stenosis or closure of the diseased artery resulting in obstructed blood flow and (b) dislodgement of arterial plaques or thromboemboli into the bloodstream, causing an obstruction distal to the lesion. Because of the sensitive anatomical location and potential risks of therapy involved with carotid arteries, therapeutic options are limited and depend on the location and extent of atherosclerosis. Current options of therapy for stenosis include carotid endarterectomy, carotid artery angioplasty, and antiplatelet pharmacological treatment.

5.8.1   Carotid Endarterectomy

Carotid endarterectomy is a surgical procedure in which the atherosclerotic or stenosed portion of the vessel wall is surgically excised.[88-92] In carotid endarterectomy, the surgeon surgically exposes the afflicted carotid artery and identifies the spatial extent of the atherosclerotic lesion. In order to, in effect, isolate the carotid artery for the surgical procedure, blood flow must be stopped through this segment and redistributed via an externally placed shunt. Since the carotid artery accounts significantly for the brain's blood supply, one must be certain that, following isolation of the carotid artery, the blood flow to the brain will maintain adequate levels of perfusion. To ensure adequate perfusion following temporary vessel obstruction required for endarterectomy, the patient is subjected to temporary balloon occlusion. The purpose of this study is to determine the role and function of collateral circulation following the temporary occlusion of the afflicted artery with balloons. The balloons are navigated endovascularly to the diseased artery and inflated for a time ranging from 10 to 30 min. While the artery is occluded, the patient is subjected to a medical imaging procedure such as SPECT, a physical exam, or an electroencephalogram (EEG) for images of cerebral blood flow and function.[93,94] If the brain is perfused adequately during temporary balloon occlusion, then the surgical procedure is deemed safe within limits and the procedure is then performed. If not, other therapeutic options are investigated.

Continuing with the discussion on carotid endarterectomy, the diseased portion of the carotid artery is excised and an atherectomy or surgical removal of the fatty deposit is performed. In most cases, upon removal of the atherosclerotic lesion, the incision is sutured followed by the careful resumption of flow through the carotid artery. Although in most cases the incision is sutured, there are several instances such as affected area and contour of the artery that dictate the need for carotid patch angioplasty. In carotid patch angioplasty, a material such as saphenous vein, Dacron, or polytetrafluoroethylene is sutured over the original excision, enlarging the endarterectomized segment and restoring normal contour to the carotid bulb.[95] Although largely unsubstantiated, it is believed that carotid patch angioplasty results in a change in the arterial cross-sectional diameter, which in turn reduces shear stress, while small increases in luminal cross section reduce stress on the arterial wall.[96] Of the various types of materials used for carotid patch angioplasty, saphenous vein is the preferred material because of its ease in handling, its antithrombotic properties or resistance to formation of clots, its size, tensile strength, and resistance to infection.[95] Biomechanical studies of the venous wall subjected to arterial conditions have shown that vein performs very well in maintaining blood flow through the carotid artery[97] although rupture of the vein graft can and does occur.[98] Carotid endarterectomy is a relatively new procedure and is the subject of long-term, multicenter studies such as the North American Symptomatic Carotid Endarterectomy Trial (NASCET).[99]

5.8.2   Carotid Artery Angioplasty

Another form of therapy in treating the atherosclerotic lesion involves a technique commonly employed in the treatment of cardiovascular disease and atherosclerotic lesions of the coronary arteries. In carotid angioplasty, a balloon is transported to the atherosclerotic lesion via a catheter and subsequently inflated. The inflated balloon opens the clogged vessel by redistributing the fatty deposits along the artery wall.[100,101] This technique is relatively new and requires additional testing on larger patient populations to determine its clinical efficacy.

5.8.3   Pharmacological Treatment

Another option of therapy available to the patient is pharamacological treatment involving thrombolysis or breaking up blood clots. Thrombolytic therapy involves the use of biochemical enzymes known as thrombolytic agents that exhibit a strong affinity to the end product of the clotting system (fibrin) and interact with the blood clotting process to lyse or break down the clot.[102,103] Dissolution of the blood clot or thrombus removes the intraluminal obstruction (recanalization) and reestablishes cerebral blood flow.

5.9   Summary

The clinical origin and causes of stroke are attributed not to one single factor but a host of factors acting independently, in combination, or in succession. These factors involve possible genetic, biochemical, biophysical, physiological, anatomical, and histological components. As a result, the genetic component to the genesis of stroke and resultant neurological consequences exerts a possible influence and, most likely, dictates the various altered outcomes of the remaining components. In addition, options available in the treatment of stroke form part of a complicated issue involving a careful consideration of the risks to the patient.[104] Furthermore, experimental studies have shown that cerebral ischemia induces a variety of changes in gene expression in the brain.[105] The subject matter presented in this chapter addressed primarily the biophysical processes implicated in stroke and, to a lesser extent, anatomical and physiological components.

5.10 References

1. American Heart Association, *Heart and Stroke Facts: 1996 Statistical Supplement* (American Heart Association, Dallas, 1996).
2. P.J. Camarata, R.C. Heros, and R.E. Latchaw, "Brain Attack: The rationale for treating stroke as a medical emergency," Neurosurgery **34**, 144–158 (1994).
3. D. Barron and S. Starkman, "Emergency evaluation and management of stroke: Part I. Hemorrhagic stroke," Hospital Physician **1995**, February, 16–25.
4. D. Barron and S. Starkman, "Emergency evaluation and management of stroke: Part II. Ischemic stroke," Hospital Physician **1995**, March, 21–49.
5. N. Woolf and M.J. Davies, "Arterial plaque and thrombus formation," Sci. Am. Sci. Med. **1**, 38–47 (1994).
6. R. Ross, "The pathogenesis of atherosclerosis: A perspective for the 1990's," Nature **362**, 801–809 (1993).
7. R. Ross, "Atherosclerosis: A problem of the biology of arterial wall cells and their interactions with blood components", Arteriosclerosis **1**, 293–311 (1981).
8. D.P. Hajjar and A.C. Nicholson, "Atherosclerosis," Am. Scientist **83**, 460–467 (1995).
9. M.S. Brown and J.L. Goldstein, "A receptor-mediated pathway for cholesterol homeostasis," Science **232**, 34–47 (1986).
10. F.B. Gessner, "Hemodynamic theories of atherogenesis," Circ. Res. **33**, 259–266 (1973).
11. D.L. Fry, "Acute vascular endothelial changes associated with increased blood velocity gradients," Circ. Res. **22**, 165–197 (1968).
12. P.D. Stein and H.N. Sabbah, "Measured turbulence and its effect on thrombus formation," Circ. Res. **35**, 608–614 (1974).
13. J.F. Mustard, E.A. Murphy, H.C. Roswell, and H.G. Downie, "Factors influencing thrombus formation in vivo," Am. J. Med. **33**, 621–647 (1962).
14. H.L. Goldsmith and V.T. Turitto "Rheological aspects of thrombosis and haemostasis: Basic principles and applications," Thromb. Haemostasis **55**, 415–435 (1986).
15. J.A. DeMatteis, "Ischemic stroke: Etiology and treatment," Hospital Physician **1994**, June, 55–64.
16. R.H. Wilkins, "Cerebral vasospasm," Contemp. Neurosurg. **10**, 1–66 (1988).
17. S. Roux, B.M. Löffler, G.A. Gray, U. Sprecher, M. Clozel, and J.-P. Clozel, "The role of endothelin in experimental cerebral vasospasm," Neurosurgery **37**, 78–86 (1995).
18. T. Shigeno, M. Clozel, S. Sakai, A. Saito, and K. Goto, "The effect of bosentan, a new potent endothelin receptor antagonist, on the pathogenesis of cerebral vasospasm," Neurosurgery **37**, 87–91 (1995).
19. M. Zimmermann, U. Seifert, B.-M. Löffler, D. Stolke, and W. Stenzel, "Pre-

vention of cerebral vasospasm after experimental subarachnoid hemorrhage by RO 47–0203, a newly developed orally active endothelin receptor antagonist," Neurosurgery **38**, 115–120 (1996).

20. J.M. Findlay, N.F. Kassell, B.K.A. Weir, E.C. Haley, Jr., G. Kongable, T. Germanson, L. Truskowski, W.M. Alves, R.O. Holness, N.W. Knuckey, H. Yonas, G.K. Steinberg, M, West, H.R. Winn, and G. Ferguson, "A randomized trial of intraoperative, intracisternal tissue plasminogen activator for the prevention of vasospasm," Neurosurgery **37**, 168–178 (1995).

21. C.W. Kerber and D. Liepsch, "Flow dynamics for radiologists. II. Practical considerations in the live human," Amer. J. Neuroradiol. **15**, 1076–1086 (1994).

22. M. Motomiya and T. Karino, "Flow patterns in the human carotid artery bifurcation," Stroke **15**, 50–55 (1984).

23. M.R. Roach, S. Scott, and G.G. Ferguson, "The hemodynamic importance of the geometry of bifurcations in the circle of Willis (glass model studies)," Stroke **3**, 255–267 (1972).

24. R.S. Salzar, M.J. Thubrikar, and R.T. Eppink, "Pressure-induced mechanical stress in the carotid artery bifurcation: A possible correlation to atherosclerosis," J. Biomech. **28**, 1333–1340 (1995).

25. M. Fisher and S. Fieman, "Geometric factors of the bifurcation in carotid atherogenesis," Stroke **21**, 267–271 (1990).

26. X.M. Pan, D. Saloner, L.M. Reilly, J.C. Bowersox, S.P. Murray, C.M. Anderson, G.A.W. Gooding, and J.H. Rapp, "Assessment of carotid artery stenosis by ultrasonography, conventional angiography, and magnetic resonance angiography: Correlation with ex vivo measurement of plaque stenosis," J. Vasc. Surg. **21**, 82–89 (1995).

27. D.F. Young, N.R. Cholvin, R.L. Kirkeeide, and A.C. Roth, "Hemodynamics of arterial stenoses at elevated flow rates," Circ. Res. **41**, 99–107 (1977).

28. A.G. May, J.A. DeWeese, and C.G. Rob, "Hemodynamic effects of arterial stenosis," Surgery **53**, 513–524 (1963).

29. R.V. Fiddian, D. Byar, and E.A. Edwards, "Factors affecting flow through a stenosed vessel," Arch. Surg. **88**, 83–90 (1964).

30. N.A. Delin, S. Ekeström, and R. Telenius, "Relation of degree of internal carotid artery stenosis to blood flow and pressure gradient: An angiographic and surgical study in man," Invest. Radiol. **3**, 337–344 (1968).

31. B.Y. Lee, C. Assadi, J.L. Madden, D. Kavner, F.S. Trainor, and W.J. McCann, "Hemodynamics of arterial stenosis," World J. Surg. **2**, 621–629 (1978).

32. J.G. Brice, D.J. Dowsett, and R.D. Lowe, "Haemodynamic effects of carotid artery stenosis," Brit. Med. J. **2**, 1363–1366 (1964).

33. D. Byar, R.V. Fiddian, M. Quereau, J.T. Hobbs, and E.A. Edwards, "The fallacy of applying the Poiseuille equation to segmental arterial stenosis," Am. Heart J. **70**, 216–224 (1965).

34. A. Roos, "Poiseuille's law and its limitations in vascular systems," Med. Thorac. **19**, 224–238 (1962).

35. A.G. May, L. Van de Berg, J.A. DeWeese, and C.G. Rob, "Critical arterial stenosis," Surgery **54**, 250–259 (1963).

36. R. Berguer and N.H.C. Hwang, "Critical arterial stenosis: A theoretical and experimental solution," Ann. Surg. **180**, 39–50 (1974).

37. J.P. Archie, Jr. and R.W. Feldtman, "Critical stenosis of the internal carotid artery," Surgery **89**, 67–72 (1981).

38. G.W. Kindt and J.R. Youmans, "The effect of stricure length on critical arterial stenosis," Surg. Gynecol. Obstet. **128**, 729–734 (1969).

39. J.A. Deweese, A.G. May, E.O. Lipchik, and C.G. Rob, "Anatomic and hemodynamic correlations in carotid artery stenosis," Stroke **1**, 149–157 (1970).

40. R.O. Bude, J.M. Rubin, J.F. Platt, K.P. Fechner, and R.S. Adler, "Pulsus tardus: Its causes and potential limitations in detection of arterial stenosis," Radiology **190**, 779–784 (1994).

41. Y.M. Akay, M. Akay, W. Welkowitz, S. Lewkowicz, and Y. Palti, "Dynamics of the sounds caused by partially occluded femoral arteries in dogs," Ann. Biomed. Eng. **22**, 493–500 (1994).

42. Y. Kurokawa, S. Abiko, and K. Watanabe, "Noninvasive detection of intracranial vascular lesions by recording blood flow sounds," Stroke **25**, 397–402 (1994).

43. D.J. Farrar, H.D. Green, M.G. Bond, W.D. Wagner, and R.A. Gobbee, "Aortic pulse wave velocity, elasticity, and composition in a nonhuman primate model of atherosclerosis," Circ. Res. **43**, 52–62 (1978).

44. A.D. Nashif, D.I.G. Jones, and J.P. Henderson, *Vibration Damping*. (Wiley, New York, 1985), p. 122.

45. M.D. Deshpande, D.P. Giddens, and R.F. Mabon, "Steady laminar flow through modelled vascular stenosis," J. Biomech. **9**, 165–174 (1976).

46. J.H. Forrester and D.F. Young, "Flow through a converging-diverging tube and its implications in occlusive vascular disease: I. Theoretical development," J. Biomech. **3**, 297–305 (1970).

47. J.H. Forrester and D.F. Young, "Flow through a converging-diverging tube and its implications in occlusive vascular disease: II. Theoretical and experimental results and their implications," J. Biomech. **3**, 307–316 (1970).

48. C. Tu, M. Deville, L. Dheur, and L. Vanderschuren, "Finite element simulation of pulsatile flow through arterial stenosis," J. Biomech. **25**, 1141–1152 (1992).

49. N. Stergiopulos, D.F. Young, and T.R. Roggee, "Computer simulation of arterial flow with applications to arterial and aortic stenoses," J. Biomech. **25**, 1477–1488 (1992).

50. B.D. Seeley and D.F. Young, "Effect of geometry on pressure losses across models of arterial stenoses," J. Biomech. **9**, 439–448 (1976).

51. N. Stergiopulos, M. Spiridon, F. Pythoud, and J.-J. Meister, "On the wave transmission and reflection properties of stenoses," J. Biomech. **29**, 31–38 (1996).

52. P.K.C. Wong, B. Eng, K.W. Johnston, C.R. Ethier, and R.S.C. Cobbold, "Computer simulation of blood flow patterns in arteries of various geometries," J. Vasc. Surg. **14**, 658–667 (1991).

53. V. Stefanovich, editor, *Stroke: Animal Models* (Pergamon, Oxford, 1983).

54. W.J. Powers, G.A. Press, R.L. Grubb, M. Gado, and M.E. Raichle, "The effect of hemodynamically significant carotid artery disease on the hemodynamic status of the cerebral circulation," Ann. Inter. Med. **106**, 27–35 (1987).

55. R. Leblanc, Y.L. Yamamoto, J.L. Tyler, and A. Hakim, "Hemodynamic and metabolic effects of extracranial carotid disease," Can. J. Neurol. Sci. **16**, 51–57 (1989).

56. W.J. Powers, "Cerebral hemodynamics in ischemic cerebrovascular disease," Ann. Neurol. **29**, 231–240 (1991).

57. M.J. Cumming and I.M. Morrow, "Carotid artery stenosis: a prospective comparison of CT angiography and conventional angiography," Amer. J. Roentgenol. **163**, 517–523 (1994).

58. M. Castillo, "Diagnosis of disease of the common carotid artery bifurcation: CT angiography vs catheter angiography," Amer. J. Roentgenol. **161**, 395–398 (1993).

59. M. Fisher J.W. Prichard, and S. Warach, "New magnetic resonance techniques for acute ischemic stroke," J. Am. Med. Assoc. **274**, 908–911 (1995).

60. M. Fisher, C.H. Sotak, K. Minematsu, and L. Li, "Innovative magnetic resonance technologies for evaluating cerebrovascular disease," Ann. Neurol. **32**, 115–122 (1992).

61. D.E. Saunders, A.G. Clifton, and M.M. Brown, "Measurement of infarct size using MRI predicts prognosis in middle cerebral artery infarction," Stroke **26**, 2272–2276 (1995).

62. M. Fisher, "Diffusion and perfusion imaging for acute stroke," Surg. Neurol. **43**, 606–609 (1995).

63. S.W. Atlas, "MR angiography in neurologic disease," Radiology **193**, 1–16 (1994).

64. M. Castillo, L. Kwock, and S.K. Mukherji, "Clinical applications of proton MR spectroscopy," Amer. J. Neuroradiol. **17**, 1–15 (1996).

65. G.D. Graham, A.M. Blamire, D.L. Rothman, L.M. Brass, P.B. Fayad, O.A.C. Petroff, and J.W. Prichard, "Early temporal variation of cerebral metabolites after human stroke: A proton magnetic resonance spectroscopy study," Stroke **24**, 1891–1896 (1993).

66. P.B. Barker *et al.*, "Acute stroke: Evaluation with serial proton MR spectroscopic imaging," Radiology **192**, 723–732 (1994).

67. J.H. Duijn, G.B. Matson, A.A. Maudsley, J.W. Hugg, and M.W. Weiner, "Human brain infarction: Proton MR spectroscopy," Radiology **183**, 711–718 (1993).

68. W.M. Blackshear *et al.*, "Detection of carotid occlusive disease by ultrasonic imaging and pulsed Doppler spectrum analysis," Surgery **86**, 698–706 (1979).

69. J.L. Glover, P.J. Bendick, V.P. Jackson, G.J. Becker, R.S. Dilley, and R.W. Holden, "Dupplex ultrasonography, digital subtraction angiography, and conventional angiography in assessing carotid atherosclerosis," Arch. Surg. **119**, 664–669 (1984).

70. G.J. Hankey, C.P. Warlow, and R.J. Sellar, "Cerebral angiographic risk in mild cerebrovascular disease," Stroke **21**, 209–222 (1990).

71. C. Raynaud, G. Rancurel, N. Tzourio, J.P. Soucy, J.C. Baron, S. Pappata, H. Cambon, B. Mazoyer, N.A. Lassen, E. Cabanis, A. Majdalani, M. Bourdoiseau, S. Ricard, M. Bourguignon, and A. Syrota, "SPECT analysis of recent cerebral infarction," Stroke **20**, 192–204 (1989).

72. J.V. Bowler, J.P.H. Wade, B.E. Jones, K. Nijran, and T.J. Steiner, "Single-photon emission computed tomography using hexamethyl propyleneamine oxime in the prognosis of acute cerebral infarction," Stroke **27**, 82–86 (1996).

73. A.E. Baird and G.A. Donnan, "Increased ^{99m}Tc–HMPAO uptake in ischemic stroke," Stroke **24**, 1261–1262 (1993).

74. R.J. English *et al.*, "Brain imaging of cerebrovascular disease with I-123 HIPDM," J. Nucl. Med. Technol. **12**, 13–15 (1984).

75. J. Hatazawa, T. Satoh, E. Shimosegawa, T. Okudera, A. Inugami, T. Ogawa, H. Fujita, K. Noguchi, I. Kanno, S. Miura, M. Murakami, H. Iida, Y. Miura, K. Uemura, "Evaluation of cerebral infarction with iodine-123-iomazenil SPECT," J. Nucl. Med. **36**, 2154–2161 (1995).

76. G. Marchal *et al.*, "PET imaging of cerebral perfusion and oxygen consumption in acute ischemic stroke: relation to outcome," Lancet **341**, 925–927 (1993).

77. I.I. Kricheff, "Arteriosclerotic ischemic cerebrovascular disease," Radiology **162**, 101–109 (1987).

78. W.J. Zweibel, "Spectrum analysis in carotid sonography," Ultrasound Med. Biol. **13**, 623 (1987).

79. J.M. De Bray P.A. Joseph, H. Jeanroine, D. Maugin, M. Dauzat, and F. Plassard, "Transcranial Doppler evaluation of middle cerebral artery stenosis," J. Ultrasound Med. **7**, 611–616 (1988).

80. R. Aaslid, P. Huber, and H. Nornes, "Evaluation of cerebrovascular spasm with transcranial Doppler ultrasound," J. Neurosurg. **60**, 37–41 (1984).

81. C.E. Withers, B.B. Gosink, A.M. Keightley, G. Casola, A.A. Lee, E. vanSonnenberg, J.F. Rothrock, and P.D. Lyden, "Duplex carotid sonography: Peak systolic velocity in quantifying internal carotid artery stenosis," J. Ultrasound Med. **9**, 345–349 (1990).

82. M. Eliasziw, R.F. Smith, N. Singh, D.W. Holdsworth, A.J. Fox, and H.J.M. Barnett, for the North American Symptomatic Carotid Endarterectomy Trial (NASCET) Group, "Further comments on the measurement of carotid stenosis from angiograms," Stroke **25**, 2445–2449 (1994).

83. P.M. Rothwell, R.J. Gibson, J. Slattery, R.J. Sellar, and C.P. Warlow, for the European Carotid Surgery Trialists' Collaborative Group, "Equivalence of measurements of carotid stenosis: A comparison of three methods on 1001 angiograms," Stroke **25**, 2435–2439 (1994).

84. P.M. Brown and K.W. Johnston, "The difficulty of quantifying the severity of carotid stenosis," Surgery **92**, 468–473 (1982).

85. A.J. Fox, "How to measure carotid stenosis," Radiology **186**, 316–318 (1993).

86. A.V. Alexandrov, C.F. Bladin, R. Maggisano, and J.W. Norris, "Measuring carotid stenosis: time for a reappraisal," Stroke **24**, 1292–1296 (1993).

87. J. Biller and B.B. Love, "Recent therapeutic options for stroke prevention," Hosp. Physician **1991**, (June), 13–24.

88. M.R. Mayberg, S.E. Wilson, F. Yatsu, D.G. Weiss, L. Messina, L.A. Hershey, C. Colling, J. Eskridge, D. Deykin, and H.R. Winn for the Veterans Affairs Cooperative Studies Program 309 Trialist Group, "Carotid endarterectomy and prevention of cerebral ischemia in symptomatic carotid stenosis," J. Am. Med. Assoc. **266**, 3289–3294 (1991).

89. R.W. Hobson, D.G. Weiss, W.S. Fields, J. Goldstone, W.S. Moore, J.B. Towne, C.B. Wright, and the Veterans Affairs Cooperative Study Group, "Efficacy of carotid endarterectomy for asymptomatic carotid stenosis," N. Engl. J. Med. **328**, 221–227 (1993).

90. C.F. Bladin, A.V. Alexandrov, and J.W. Norris, "Carotid endarterectomy and the measurement of stenosis," Stroke **25**, 709–710 (1994).

91. North American Symptomatic Carotid Endarterectomy Trial Collaborators, "Beneficial effect of carotid endarterectomy in symptomatic patients with high-grade carotid stenosis," N. Engl. J. Med. **325**, 445–453 (1991).

92. R.D. Cebul and J.F. Whisnant, "Carotid endarterectomy," Ann. Intern. Med. **111**, 660–670 (1989).

93. M. Lorberboym, N. Pandit, J. Machac, V. Holan, M. Sacher, D. Segal, and C. Sen, "Brain perfusion imaging during preoperative temporary balloon occlusion of the internal carotid artery," J. Nucl. Med. **37**, 415–419 (1996).

94. F. Tanaka, S. Nishizawa, Y. Yonekura, N. Sadato, K. Ishizu, H. Okazawa, N. Tamaki, I. Nakahara, W. Taki, and J. Konishi, "Changes in cerebral blood fow induced by balloon test occlusion of the internal carotid artery under hypotension," Eur. J. Nucl. Med. **22**, 1268–1273 (1995).

95. R.F. Spetzler, J.E. Bailes, and P.J. Apostolides, "Rationale and protocol for microsurgical carotid endarterectomy," *Microsurgical Carotid Endarterectomy*, edited by J.E. Bailes and R.F. Spetzler (Lippincott-Raven, Philadelphia, 1996), Chap. 7, pp. 105–140.

96. J.R. Robinson, Jr. and J.E. Bailes "Epidemiology, natural history, and controversies in the treatment of carotid stenosis," in *Microsurgical Carotid Endarterectomy*, edited by J.E. Bailes and R.F. Spectzler (Lippincott-Raven, Philadelphia, 1996), Chap. 1, pp. 1–23.

97. S.A. Berceli, D.P. Showalter, R.A. Sheppeck, W.A. Mandarino, and H.S. Borovetz, "Biomechanics of the venous wall under simulated arterial conditons," J. Biomech. **23**, 985–989 (1990).

98. D.L. Donovan, S.P. Schmidt, S.P. Townshend, G.O. Njus, and W.V. Sharp, "Material and structural characterization of human saphenous vein," J. Vasc. Surg. **12**, 531–537 (1990).

99. North American Symptomatic Carotid Endarterectomy Trial (NASCET) Steering Committee, "North American Symptomatic Carotid Endarterectomy Trial: Methods, patient characteristics, and progress," Stroke **22**, 711–720 (1991).

100. J.D. McKenzie, R.C. Wallace, B.L. Dean, R.A. Flom, and M.H. Khayata, "Preliminary results of intracranial angioplasty for vascular stenosis caused by atherosclerosis and vasculitis," Amer. J. Neuroradiol. **17**, 263–268 (1996).

101. M.P. Spearman, C.A. Jungreis, and L.R. Wechsler, "Angioplasty of the occluded internal carotid artery," Am. J. Neurol. Res. **16**, 1791–1796 (1995).

102. A.C. Macabasco and J.L. Hickman, "Thrombolytic therapy for brain attack," J. Neurosci. Nurs. **27**, 138–151 (1995).

103. C.H. Millikan, "Role of anticoagulants in the treatment of cerebrovascular disease," Am. J. Med. **33**, 731–737 (1962).

104. H.P. Adams, T.G. Brott, R.M. Crowell, A.J. Furlan, C.R. Gomez, J. Grotta, C.M. Helgason, J.R. Marler, R.F. Woolson, J.A. Zivin, W. Feinberg, and M. Mayberg, "Guidelines for the management of patients with acute ischemic stroke," Stroke **25**, 1901–1914 (1994).

105. K. Kogure and H. Kato, "Altered gene expression in cerebral ischemia," Stroke **24**, 2121–2127 (1993).

5.11 Problems

5.1. What is a source of error of carotid artery stenosis measurements?

5.2. The intensity of an ultrasonic wave is related to the square of the pressure wave amplitude by $I = P^2/2\rho c$, where ρ is the density of the medium and c is the velocity of sound in the medium. If the pressure wave is given by $P_x = P_0 e^{-\mu x}$, where μ is the amplitude attenuation coefficient, what is the intensity at any point x?

5.3. For an acoustic wave incident normal or perpendicular to the object, representing the ideal case in ultrasonic imaging, how do the ratios A_r/A_i and A_t/A_i change?

5.4. How are ultrasonic waves different from x rays?

5.5. For a twofold increase in the pressure drop across a stenosis, what is the corresponding increase in the rate of volumetric blood flow?

5.6. The average blood flow velocity in a blood vessel of radius $300\,\mu m$ is $100\,cm/s$ but is increased over a luminal obstruction ($R_{obs} = \frac{1}{3} R_{orig}$) to $200\,cm/s$. How much work is done by the blood as it passes through the obstructed region? (Note $\rho_{blood} = 1000\,kg/m^3$.)

5.7. A blood vessel of radius $10\,mm$ is partially obstructed to $\frac{1}{2}$ of its original radius. If the average blood speed is initially $100\,cm/s$, what is the average blood speed through the obstructed region?

6
The Physics of Intracranial Aneurysms

6.1 Introduction

Chapter 6 presents a description of the main biophysical principles, interactions, and phenomena related to the development and natural history, diagnosis, and therapy of intracranial aneurysms. The primary objectives of this chapter are to provide the reader with knowledge of (1) the clinical epidemiology and prevalence of intracranial aneurysms; (2) the various types and associated general features of intracranial aneurysms; (3) the scientific or biomathematical basis for the initiation, development, and rupture of intracranial saccular and fusiform aneurysms; (4) the theoretical and experimental methods used to investigate the pathophysiology of intracranial aneurysms; and (5) a review of techniques for the diagnosis and therapy of intracranial aneurysms.

An intracranial aneurysm (often used interchangeably with the term *cerebral aneurysm*), shown in Fig. 6.1, is a form of cerebrovascular disease that manifests itself as a pouching or ballooning of the vessel wall. Although aneurysms may develop within any area of the human body, i.e., cardiovasculature, aorta, pulmonary vasculature, and peripheral vasculature, and along any vessel of the brain, i.e., artery or vein, aneurysms are fairly common within the brain and are much more likely to occur along the course of an artery. The dilatation develops at a diseased site along the arterial wall into a distended sac of stressed thinned-out arterial tissue. The fully developed cerebral aneurysm typically ranges in size from a few millimeters to 15 mm, but can attain sizes greater than 2.5 cm. If left untreated, the aneurysm may continue to expand until it ruptures, causing subarachnoid hemorrhage, severe neurological complications, and possibly death. Hemorrhage refers to bleeding from a blood vessel or aneurysm and can occur at various locations within the human brain.

Subarachnoid hemorrhage, the most common type of hemorrhage due to ruptured aneurysms, occurs in the region of the brain known as the subarachnoid space. The two primary types of subarachnoid hemorrhage are related to their cause. Traumatic subarachnoid hemorrhage occurs as a

166

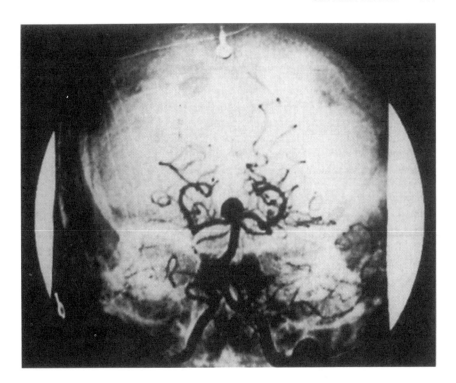

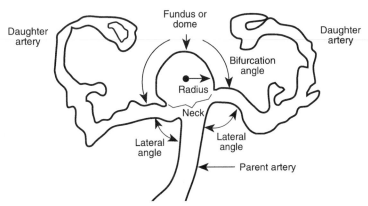

FIGURE 6.1. Top panel: Line drawing illustrating the geometric features of a human intracranial saccular aneurysm occurring at the arterial bifurcation. Bottom panel: Angiographic projectional image of a human intracranial saccular aneurysm.

direct result of injury to the brain (e.g., in motor vehicle accidents), while nontraumatic subarachnoid hemorrhage occurs without the presence of trauma. Nontraumatic subarachnoid hemorrhage due to the rupture of saccular aneurysms accounts for 75% of all documented cases.[1] In the United States, approximately 20 000 aneurysms rupture each year; 8% of these patients die before reaching a hospital.[2] About 50% of patients with ruptured aneurysms die or become permanently disabled as a result of the initial hemorrhage, and another 25%–35% die of a future hemorrhage. Thus, only 12%–15% of patients who sustain a ruptured intracranial aneurysm will survive the episode relatively unscathed.[3] Severe neurological complications refer to deficits or limitations experienced by the patient with regard to normal brain function and processes. These complications are typically measured in terms of a patient's ability to adequately perform daily activities. Scientific and clinical research into the origin, development, and rupture of cerebral aneurysms are inconclusive and theories deduced to adequately explain the stages of aneurysm development are, at best, speculative.[4] A thorough and comprehensive understanding of these developmental stages is necessary to effectively diagnose and treat intracranial aneurysms and improve patient outcome.

6.2   Natural History of Intracranial Aneurysms

Clinical assessment and therapy of intracranial aneurysms are usually based on the natural history of the aneurysm.[5-9] The natural history represents a compendium of documented information describing the progression of the disease and consists primarily of clinical observations, results from genetic, biochemical, histological, biophysical, and biomechanical experimentation, and presentation from diagnostic images. For example, intracranial aneurysms are highly likely at irregular arterial geometries such as tortuous segments, bends, and bifurcations. Bifurcations are points at which one artery subdivides into two arteries and resemble "forks in the road." In fact, 90% of the aneurysms encountered are berry or saccular aneurysms, which occur most often at points of bifurcation along the large arteries that constitute the circle of Willis and its proximal branches.[1,10] Located at the base of the brain (refer to Fig. 3.7), the circle of Willis is a complex arterial network in which all of the major cerebral arteries converge in a pentagonal arrangement before branching into various configurations and directions. The location of these aneurysms as they pertain to the major vessels within the circle of Willis is depicted in Fig. 6.2. Although the natural history of intracranial aneurysms is generally accepted among neurologists, neurosurgeons, and neuroradiologists, controversy still exists on a number of critical issues such as the pathogenesis and prevalence of aneurysms, patient characteristics and risk factors known to correlate with the incidence of aneurysms, the critical size of the aneurysm at rupture, pre- and post-operative management of aneurysms, and selection of therapy for a particular aneurysm.

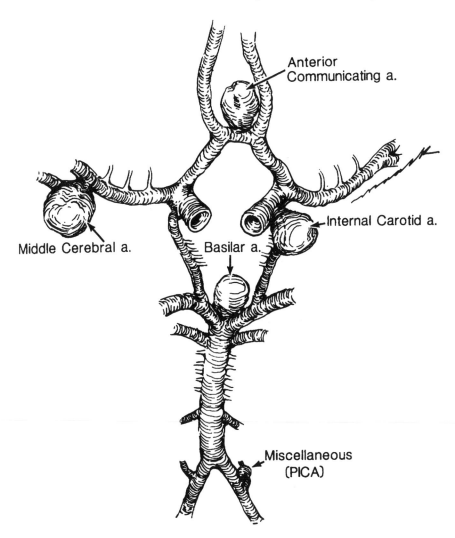

FIGURE 6.2. Schematic diagram of the frequent appearance of intracranial aneurysms within the circle of Willis. (Source: A.G. Osborn, Intracranial aneurysms: clinico-pathologic correlations. In: *Core Curriculum Course in Neuroradiology. Part I: Vascular Lesions and Degenerative Diseases*, edited by M.N. Brant-Zawadzki, B.P. Drayer, p. 7. © 1995 by American Society of Neuroradilogy.)

Currently, the origin and development of intracranial aneurysms are be-lieved to be attributed to a variety of factors classified as either congenital or acquired.[4] Congenital factors implicate a genetic abnormality known as a medial defect or localized area within the artery that lacks the power of contraction.[11] The congenital nature of intracranial aneurysms has prompted investigations into the incidence of aneurysms transmitted through family generations[12–16] and the development of a screening protocol for patients with such a family background.[17,18] If one has a family history of aneurysms, the patient is said to be genetically predisposed or at a higher risk for an aneurysm than those who have no family history due to the increased probability that the patient carries the defective gene coding for the occurrence of an aneurysm. As an interesting sidenote to their congenital nature, intracranial aneurysms are rarely observed at birth or early infancy[19] but have been routinely observed in children and adolescents.[20,21] Although theories based on genetic predisposition are widely believed to play an integral role in aneurysm initiation and development, they have not been substantiated with evidence other than clinical observations and, until a gene unique to weaknesses in the arterial wall and the definite presence of aneurysms has been identified and isolated, these ideas cannot be considered as anything other than a hypothesis.[4,22] Possible acquired factors involved in the pathogenesis and development of aneurysms include (1) injury to the endothelial wall caused by the oscillatory shear stresses of blood flow,[23,24] (2) enzymatic destruction of the vessel wall,[25] (3) immature collagen,[26] and (4) collagen deficiency.[27–29] The term *endothelial* pertains to the layer of cells which line the inner wall of the blood vessel, as described in Chap. 3. Since blood is pumped from the heart on a periodic basis by alternate dilation and contraction, the flow of blood proceeds in a rhythmic or pulsatile fashion. The behavior of pulsatile blood flow mimics that of an oscillation or oscillatory function. This is in contrast to constant flow where fluid flow occurs as a continuous stream. The presence of flow, whether constant or pulsatile, induces shear stresses that are at a maximum along the inner wall of the vessel, as discussed in Chap. 4.

Enzymatic destruction of the vessel wall occurs as the result of enzymes or biomolecules that act as catalysts in accelerating the rate of chemical or biological reactions of a substance for which it is specific. In this case, enzymes such as collagenase and elastase increase the rate of degradation of the structural components of the vessel wall, collagen and elastin, respectively. Collagen and elastin are biomolecules responsible for the structural integrity and elasticity of the blood vessel and are typically found in connective tissues, tendon, bone, and cartilage (see Chap. 3). One way to describe quantitatively the stability of collagen is by Young's modulus, which represents the elastic resistance to tangential and shearing forces. The strength of collagen is apparent from its large value of Young's modulus of 1×10^8 dyn/cm^2. Elastin is a weaker biomolecule with a Young's modulus of 5×10^6 dyn/cm^2 but also plays an important role in vessel structure. Therefore, it stands to reason that any defect or deficiency in the metabolism of collagen or elastin would considerably weaken the vessel wall, making it structurally unstable and susceptible to aneurysm development. Hypertension,[30–32] atherosclerosis,[9,33] inflammation,[34] and connective tissue disorders[35] associated with acquired loss of tensile strength of the connective tissues have also been identified as possible causative factors. In all

Color Plates

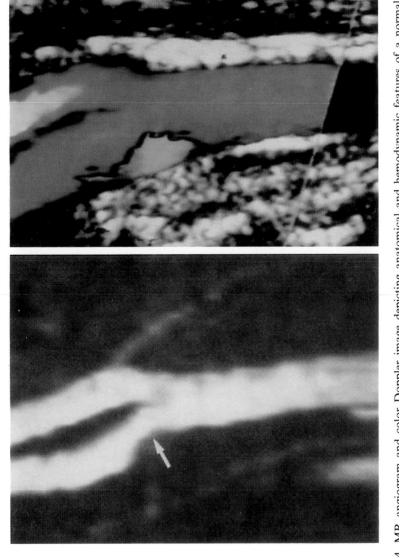

FIGURE 5.4. MR angiogram and color Doppler image depicting anatomical and hemodynamic features of a normal carotid bifurcation. (Reprinted, with permission, from *Radiology* **185**, R. L. Wolf, D. B. Richardson, C. C. LaPlante, J. Huston, III, S. J. Riederer, and R.L. Ehman, Blood flow imaging through detection of temporal variations in magnetization, pp. 559–567. © 1992 by the Radiological Society of North America.)

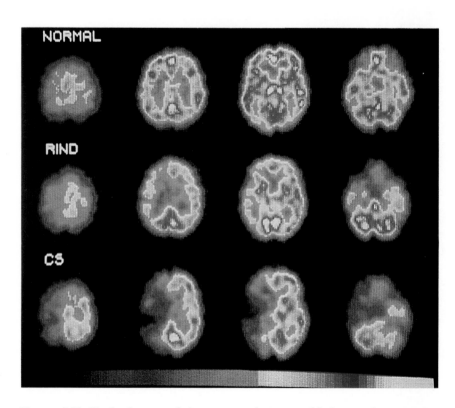

FIGURE 5.11. Single-photon emission computed tomographic images that represent various stages of cerebral ischemia as indicated by the reduction in cerebral blood flow. RIND represents reversible ischemic neurological deficit; CS, completed stroke. (Reproduced, with permission, from *Stroke*, **20**, I. Podreka, C. Baumgartner, E. Suess, C. Müller, T. Brücke, W. Lang, F. Holzner, M. Steiner, and L. Deecke, Quantification of regional cerebral blood flow with IMP-SPECT: Reproducibility and clinical relevance of flow values, pp. 183–191. © 1989 American Heart Association.)

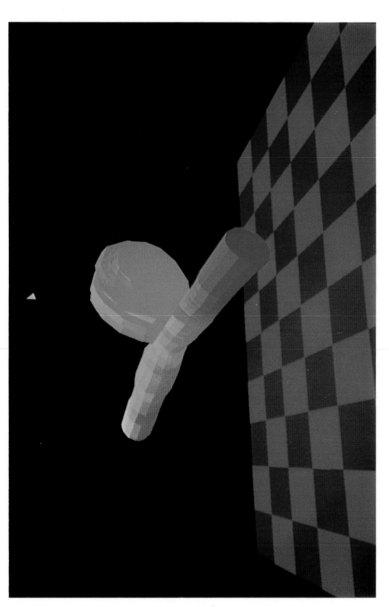

FIGURE 6.21. Three-dimensional image of an experimentally produced cerebral aneurysm of a swine model reconstructed from a sequence of digital subtraction angiography projection radiographs and displayed in a virtual reality environment. The dimension provided by the colors reveals the relative pressure distribution determined from numerical simulation. Red represents maximum values of pressure while blue represents minimum values of pressure. (Reprinted, with permission, from G.J. Hademenos, The physics of cerebral aneurysms, *Physics Today* **48**, 24–30, 1995.)

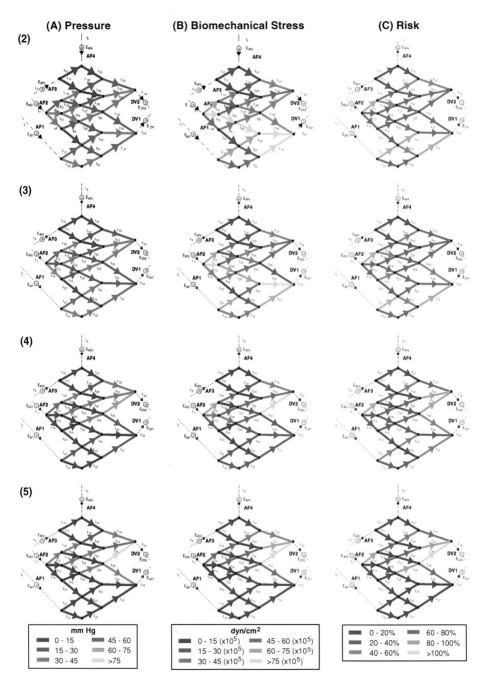

FIGURE 7.5. Schematic diagram of the nidus portion of the biomathematical AVM model depicting the intranidal values of (A) intravascular pressure gradient, (B) biomechanical stress, and (C) risk of rupture with draining vein (DV) 1 occluded 25%, 50%, 75%, and 100% and DV2 patent. The number along the left-hand column refers to the simulation number. (Reproduced, with permission, from *Stroke, 27*, G.J. Hademenos and T.F. Massoud, Risk of intracranial arteriovenous malformation rupture due to venous drainage impairment: A theoretical analysis, pp. 1072–1083. © 1996 American Heart Association.)

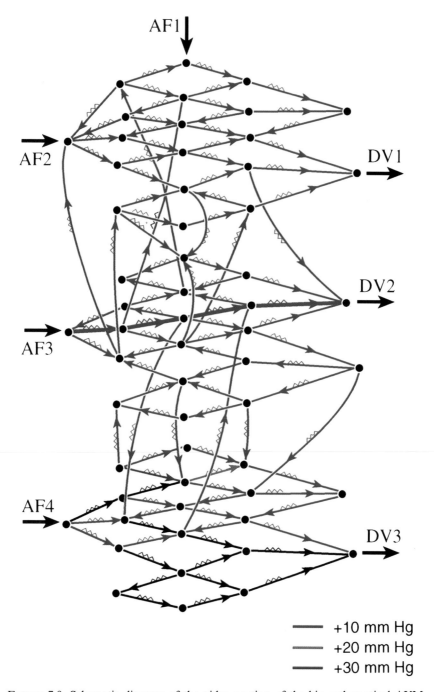

FIGURE 7.9. Schematic diagram of the nidus portion of the biomathematical AVM model simulating SSA through arterial feeder (AF) 3 at injection pressures of 10, 20, and 30 mm Hg represented by the color shaded regions.

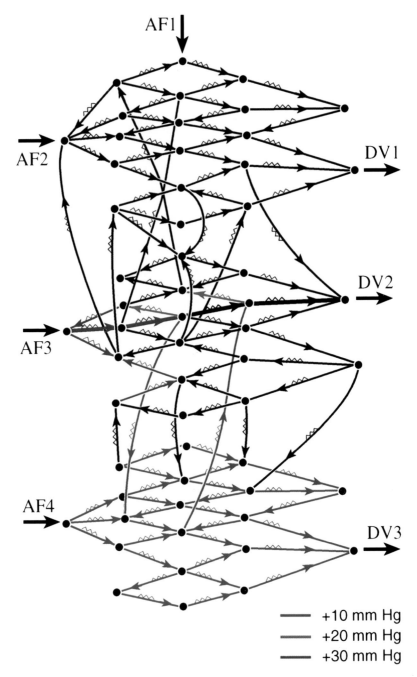

FIGURE 7.10. Schematic diagram of the nidus portion of the biomathematical AVM model simulating SSA through arterial feeder (AF) 4 at injection pressures of 10, 20, and 30 mm Hg represented by the color shaded regions.

probability, each of these factors, both congenital and acquired, contributes to some extent to the development of aneurysms.

Present knowledge of cerebral saccular aneurysms is limited to sites of occurrence,[36,37] biophysical interactions,[38–41] histological or microscopic study of aneurysm tissue structure,[42–45] radiographic identification,[8,46,47] structural and anatomical changes of the brain induced by aneurysm rupture,[48,49] and endovascular or neurosurgical treatment.[50] Endovascular and neurosurgical techniques represent two methods for treating aneurysms and differ primarily in their approach for accessing the aneurysm to administer therapy. Neurosurgery involves direct entrance into the cranial cavity and exposing the portion of brain containing the aneurysm. The aneurysm is then secured and clamped at its base with a surgical clip. In endovascular techniques, delivery of therapy is accomplished through catheters or long, thin, flexible, biocompatible tubes that are threaded through the arteries in the body up to the aneurysm. Through the hollow catheter, mechanical agents such as coils or balloons are guided into the aneurysm for packing it and preventing the aneurysm wall from further damage. Aneurysm therapies will be covered in more detail later in the chapter.

Progress in understanding the processes involved in the initiation, growth, and rupture of aneurysms is hampered by several factors. The first factor is the complex *in vivo* nature of the problem. With respect to diagnosis, there are, in effect, two types of aneurysms: asymptomatic and symptomatic. Asymptomatic aneurysms do not produce symptoms in the patient and, unfortunately, are discovered typically on an incidental basis when performing imaging studies for other medical symptoms or once they have ruptured.[51–53] This type of aneurysm can also be incidentally discovered in the pursuit of a diagnosis for totally unrelated symptoms. Asymptomatic aneurysms occur most frequently. An unruptured but symptomatic aneurysm can come to clinical attention in a variety of ways including mass effect, headaches, seizures, or as a "fellow traveler" with a ruptured aneurysm.[54] Seizures are convulsions thought to be due to an imbalance or sudden abnormal surge of electrical impulses produced by the brain. Ruptured aneurysms are also symptomatic where the ensuing hemorrhage from the ruptured aneurysm may create a substantial increase in intracranial pressure[55] accompanied by the immediate onset of severe headaches and may be followed by unconsciousness. Also, patients with a family history of intracranial aneurysms are believed to be at higher risk for an aneurysm and are screened routinely on a systematic basis. Given these diagnostic types of aneurysms, it is virtually impossible to predict accurately the occurrence of an aneurysm and thus study initiation and early development. Once an aneurysm has been positively identified based on a radiographic diagnosis, clinical observations of size, shape or morphology, volume, and location can easily be made and therapeutic strategies are considered immediately and implemented aggressively.

The second factor involves the varying size, location, neck or lumen diameter, geometry, and morphology of the aneurysm. Each diagnosed

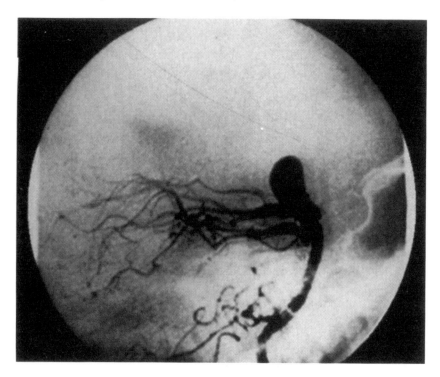

FIGURE 6.3. Angiographic images representative of the various types of aneurysm morphologies typically seen in the human population. Left panel: Oval aneurysm. Right panel: Multilobulated aneurysm. (Source: G.J. Hademenos, T.F. Massoud, and F. Viñuela. Quantitation of intracranial aneurysm neck size from diagnostic angiograms based on a biomathematical model. *Neurological Research* **17**, 322–328 1995. Reproduced by permission of the publishers Forefront Publishing.)

aneurysm is unique for a given patient and varies according to location, morphology, radius, and neck diameter. Secondary factors believed to be involved in aneurysm development include age, sex, blood pressure, patient history, hypertension, diabetes, and obesity.[56] Aneurysms are described and depicted usually as uniform spherical structures attached symmetrically to its parent vessel. Although aneurysms do present themselves as round or spherical, their morphology is typically oval or multilobulated. Multi-lobulated aneurysms are aneurysms having numerous lobules or the geo-metrical appearance of fused sacs or spheres. An example of each of these aneurysm morphologies is shown in Fig. 6.3. Compare the multilobulated and oval aneurysms in Fig. 6.3 with a round aneurysm presented in Fig. 6.1. Aneurysm sizes are classified according to the following classification of diameters (D): small, $D < 12.0$ mm; large, $12.0\,\text{mm} \leq D \leq 25.0\,\text{mm}$; giant, $D > 25.0\,\text{mm}$. Even though aneurysms typically range in size from 2 to 20 mm in diameter and tend to rupture at sizes equal to or greater than

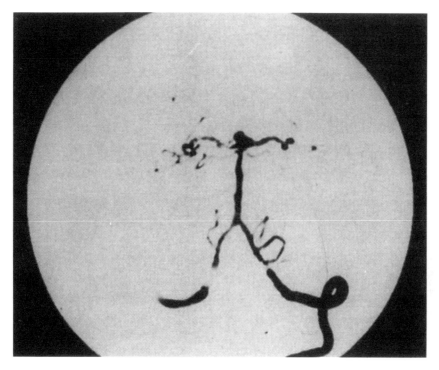

FIGURE 6.3 (continued)

10 mm in diameter,[6] they have been documented to reach sizes as large as 2.5 cm in diameter or larger. One would expect that aneurysms stretched to sizes of this magnitude would exceed their elastic limit and rupture. It has been observed, however, that these large aneurysms, referred to in the literature as *giant aneurysms*,[57,58] tend to promote intra-aneurysmal thrombosis and subsequently reduce circumferential stress in the aneurysm and thus the probability of rupture.[59-61] Also, Ferguson, Peerless, and Drake[62] have presented clinical evidence in support of the fact that rupture is possible in any aneurysm, regardless of size.

The final factor is the lack of correlated data between aneurysm parameters and physiological variables. It could be deduced logically that those experiencing either sudden increases in blood pressure or chronic high blood pressure (hypertension) would be at greater risk for aneurysm rupture than those with normal blood pressure. Jain[63] was one of the first to postulate that a possible factor in aneurysm rupture was a resonant frequency induced by pulsatile blood flow. This hypothesis was based on clinical observations of bruits or high-pitched tones heard within the aneurysms. Quantitative information between the resonant frequency of the aneurysm wall, the pulsatile frequency of blood flow, or relations between these two frequencies and physiological parameters such as heart rate and blood pressure remain unknown. Locksley[64] evaluated clinically the normal daily activities of a large number of patients with documented cases of aneurysm rupture and concluded that rupture was just as probable during sleep as during strenuous exercise, i.e., there was no influence of pulsatile frequency on rupture. This was further substantiated by the findings from Freytag.[65]

6.3   Experimental Models of Intracranial Aneurysms

Their unpredictable nature makes the study of aneurysms *in vivo* extremely difficult. Since the image resolution of current radiographic technology is not capable of revealing information regarding the tissue structure of the aneurysm wall, *in vivo* analysis of cerebral aneurysms is limited primarily to clinical observations in patient studies and tissue specimens retrieved at autopsy. However, a variety of *in vitro* methods exist for characterizing and studying the development and rupture of intracranial aneurysms. These methods include (1) biomathematical models of the aneurysm based on biophysical principles and physiological phenomena[66-72]; (2) numerical modeling and simulation of blood flow through a cylindrical artery structured with a geometric representation of an aneurysm[38,73-78]; (3) *in vitro* experiments of blood flow through transparent glass and plastic elastic arteries constructed or fitted with aneurysms of size, location, and geometry typically found in clinical cases;[35,57,79-84] and (4) experimentally inducing or surgically creating aneurysm models in various species of laboratory animals.[85-88] Each method, individually, allows one to study the physical or biological behavior of the aneurysm under controlled conditions.

Each of these methods are useful as a collective means of providing (1) a systematic and effective way of assembling existing knowledge about an aneurysm; (2) identification of important parameters and determination of the overall sensitivity to variation in each parameter involved in aneurysm growth and rupture; (3) calculation of quantitative values of variables that are difficult or impossible to measure; (4) a method to test hypotheses rapidly, efficiently, and inexpensively; (5) identification of specific elements or information gaps that must be further quantified, thus leading to the development of important experiments or quantitative measures; and (6) an effective model which can be utilized to predict the behavior of a true aneurysm.[89]

6.3.1   Biomathematical Models

Biomathematical models provide a solid foundation for the investigation of the influence and interrelationships between the biophysical principles and physiological phenomena involved in the initiation, development, and rupture of intracranial aneurysms. Fig. 6.4 displays a typical diagram in the description of biomathematical models of aneurysms. These models are de-

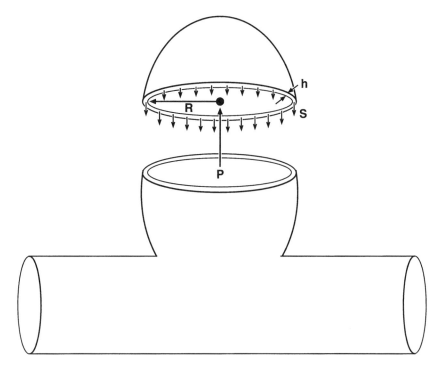

FIGURE 6.4. A graphical representation of a biomathematical model for the development of intracranial aneurysms using a free body diagram to depict the forces acting on an aneurysm. P represents pressure; R, aneurysm radius; σ, wall stress; h, wall thickness.

rived from mathematical descriptions of the statics and dynamics of the aneurysm and the biophysical response to external forces such as the hemodynamic forces or the forces produced by the flowing blood and the mechanical load placed on the aneurysm by vascular tethering. When discussing the flow of blood, it is tempting to consider arteries and veins as isolated vessels, which is actually a major assumption in the derivation and explanation of the biophysical equations and phenomena presented in the text. Although this assumption is sufficient for most calculations, it should be stated that the vessels are, in effect, permanently attached to the external or surrounding tissue.

Since the aneurysm is, in essence, an elastic sphere, we can describe the statics of the aneurysm by Laplace's law and Hooke's law, as previously described in Chap. 2. Laplace's law describes the circumferential stress along the walls of an elastic sphere and relates this to the radius of the sphere and the pressure within the cavity. Hooke's law represents the fundamental relationship between the stress and strain of an elastic object. These classical laws of physics tend to describe the aneurysm on more of a qualitative than quantitative basis but are, nevertheless, useful to the study of intracranial aneurysms. The structure of an aneurysm deviates from that assumed in Laplace's law in that the aneurysm (1) is blood-filled, (2) exhibits an irregular wall surface; (3) presents according to different morphologies; (4) contains a neck or lumen resembling a truncated sphere; (5) is often present with blebs or localized distensions in the aneurysm wall; (6) wall is composed of distinct layers of collagen, elastin, smooth muscle cells, and endothelial cells, all of which contribute to some extent to its overall elasticity; and (7) exhibits a nonlinear, viscoelastic behavior.[90]

The study of the dynamics of an aneurysm involves an investigation of the vibrational displacements of the elastic aneurysm wall as a result of the pulsatile hemodynamic forces. Elastodynamic theory has addressed, characterized, and derived the mathematical equations that describe the response of a linearly elastic and even viscoelastic sphere to external oscillating forces. The mathematics involved in the derivation of such relations as well as proper solutions is considerably complex and beyond the scope of this book. However, the interested reader is referred to Ref. 91 for further reading. Another approach in studying the dynamics of aneurysms is to determine and equate systematically the forces acting on the system in terms of parameters presented by the spherical aneurysm. Solution of such an equation would provide a time-dependent expression for the displacement of the aneurysm wall in response to the hemodynamic forces. A more detailed explanation of each of these applications will be given later in the chapter.

6.3.2 Computational Simulations

Computer simulations of aneurysm hemodynamics are performed using software packages designed for computational fluid dynamics.[92–94] The flexibility in geometry and boundary definitions allows one to construct elaborate geometrical structures very similar to true aneurysms. Using finite-element techniques, a mesh or a computer-generated representation of the geometry through which flow will traverse is created that provides the mathematical framework for the spatial and temporal flow calculations [see Fig. 6.5(A)]. Finite-element techniques represents a mathematical approach for the solution of complex differential equations subject to complicated boundary conditions. A proper discussion of finite-element techniques is beyond the scope of this book. The reader is referred to Ref. 95 for additional reading. Spatial propagation of the fluid is determined by solution of Navier–Stokes equations subject to predefined boundary conditions at finite time intervals for every point within the geometry. The Navier–Stokes equation, defined by

$$\frac{\partial \mathbf{v}}{\partial t} + (\mathbf{v} \cdot \mathbf{V})\mathbf{v} = -\frac{1}{\rho}\mathbf{V}P + \nu\mathbf{V}^2\mathbf{v},$$

where $\mathbf{v}$ is the velocity vector, ρ is the density of blood, P is the pressure, and ν

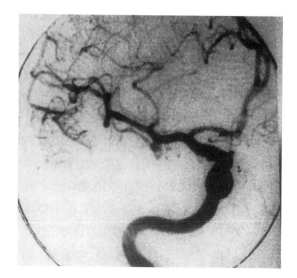

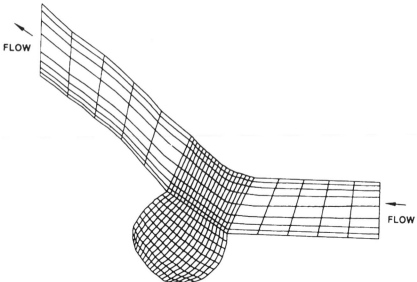

FLOW

FLOW

FIGURE 6.5. (A, p. 177) Computer-generated mesh from angiographic features of an intracranial aneurysm used in the computational simulation of hemodynamics. (Source: C. F. Gonzalez, Y. I. Cho, H. V. Ortega, and J. Moret, Intracranial aneurysms: flow analysis of their origin and progression, *American Journal of Neuroradiology* **13**, 181–188, 1992. © by American Society of Neuroradiology.) (B, pp. 178–179) Photographs showing intra-aneurysmal hemodynamics using an in vitro aneurysm model. The colored dyes are used to visualize the hemodynamic patterns. (Source: II. C.W. Kerber and D. Liepsch, Flow dynamics for radiologists. II. Practical considerations in the live human. *American Journal of Neuroradiology* **15**, 1076–1086, 1994. © by American Society of Neuroradiology.)

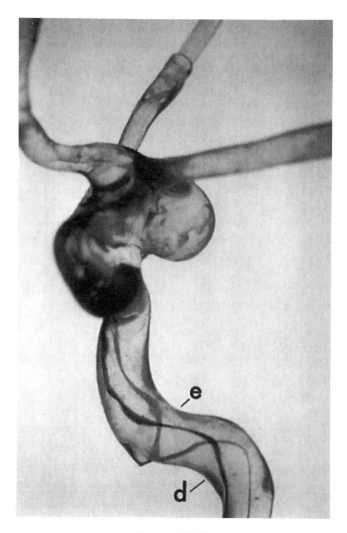

FIGURE 6.5 (B)

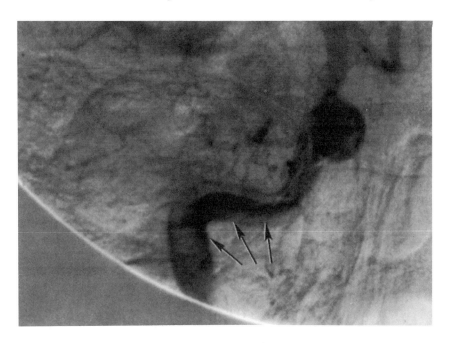

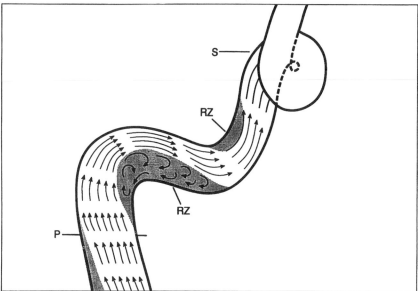

FIGURE 6.5 (B, *continued*)

is the kinematic viscosity, describes the three-dimensional motion of a fluid. With the appropriate initial and boundary conditions, the velocity and pressure can be calculated at any defined time. The simulations are greatly simplified by linearization of the Navier–Stokes equation and omitting nonlinear terms. This also reduces substantially the time to complete a simulation. As the calculations are performed at each time step, motion of the fluid (direction and magnitude) can be visualized as a dynamic sequence or movie. Three-dimensional visualization of the flow rate and other hemodynamic parameters such as velocity, pressure, shear stress, and turbulence are distinct advantages of computer simulations. Computational simulations can effectively be used to investigate intra-aneurysmal hemodynamics and transitions to turbulence within a three-dimensional geometric model of an aneurysm.

6.3.3   In Vitro *Vascular Phantoms*

In vitro vascular phantoms are three-dimensional models depicting a physical or mechanical representation of an aneurysm and attached vasculature and are used to facilitate the experimental investigation of hemodynamics within an aneurysm. Experiments involving an *in vitro* model of an aneurysm consists primarily of three individual components: aneurysm phantom, fluid, and a pump. Aneurysm phantoms are constructed primarily from glass, which provides a three-dimensional rigid geometry. More elaborate phantoms can be constructed by injecting acrylics into a postmortem sample and creating a mold.[83] Acrylics are organic substances that rapidly polymerize or solidify into a hardened plastic. The elastic phantom is difficult to manage and requires specialized equipment. In addition, the phantoms, regardless of whether they are constructed from glass or plastic, are transparent. This allows one to visualize intra-aneurysmal hemodynamics either by using a contrast agent and imaging by x-ray radiography or by using colored dyes for visualization with the human eye. Blood flow within an *in vitro* aneurysm phantom, as depicted by the streamlines from the various colored dyes, are shown in Fig. 6.5(B). The second component integral to phantom experiments is the fluid. A contrast agent is a compound that, when administered into the bloodstream during the x-ray imaging procedure, allows one to differentiate visually the high contrast of vessels containing the contrast agent on the x-ray radiograph. A variety of fluids can be used for flow experimentation depending on the level of desired realism. Water is the most commonly used fluid in such experiments and is easily and readily available. More complex solutions exist that resemble blood more closely in its physical state but are difficult to maintain and much more expensive and can permanently damage flow equipment. The final component is the pump, which controls delivery of the fluid through the vascular phantom. The most elementary type of pump is the standard water faucet that provides the phantom with a continual stream of water. However, since the circulatory system pumps blood in cycles, pulsatile pumps can be used. Pulsatile pumps allow one to control the amplitude, frequency, time, and stroke volume for each pulse of fluid ejected by the pump through the phantom. *In vitro* vascular phantoms permit the investigation and visualization of pulsatile hemodynamics within a realistic three-dimensional geometry although the individual components involved in the development of *in vitro* phantom experiments may be quite expensive.

6.3.4 Animal Models

The natural occurrence of aneurysms in animals is extremely rare. The rarity of aneurysms in lower animals has been attributed to their shorter life span, the narrower caliber of their arteries, and the lesser severity of atherosclerosis in species other than man.[96] However, experimental aneurysms can be created in animals for studies into the pathophysiology and therapy of aneurysms. Experimental aneurysms in animals may provide a realistic model of an aneurysm from a structural standpoint and have been created successfully in laboratory rats,[97,98] swine,[87,88] dogs,[99,100] rabbits,[101] and monkeys.[102] Experimental aneurysm models are typically constructed in the following four ways[87]: (1) induction of medial necrosis by injection or incubation with noxious substances[103,104]; (2) combination of carotid occlusion, induced hypertension, and induced lathyrism or chemical degradation of the vessel wall architecture[102]; (3) the use of laser energy to seal experimentally created arteriotomies (surgical removal of artery segments)[105]; and (4) surgically created aneurysms, most commonly by implantation of a venous pouch onto an artery. Yet another type of animal model is the mottled mouse, which is a mutant mouse prone to aneurysms primarily due to abnormal cellular copper metabolism.[106] Experimental animal aneurysms can be similar to human aneurysms from a morphological standpoint, allow the investigation of pulsatile *in vivo* intra-aneurysmal hemodynamics, and provide an effective means for the testing of therapeutic procedures. Several disadvantages of animal models are the requirement of surgical techniques, unrepresentative histology, reluctance to rupture, and their remoteness from sensitive neuronal tissue.

These experimental aneurysm models are generally based on the following assumptions.

1. *Ideal spherical geometry.* In most cases, saccular aneurysms exhibit a spherical appearance and can be accurately modeled by a spherical shell. However, pathological examination of a developing aneurysm may depict a locus or obtrusion[4,41] extending from the fundus or dome of the aneurysm that often occurs at the site of maximum hemodynamic stress and tends to distort the spherical symmetry. The loculi are localized regions of weakened tissue that creates also a nonuniform distribution of elasticity along the interior wall. Although aneurysms are treated typically as spherical structures, they often present themselves as oval and multilobulated morphologies as well. Although experimental *in vivo* aneurysm models can be constructed to represent oval as well as round aneurysms, they serve only as a geometric representation of an aneurysm, do not exhibit the characteristic histo-

pathology of an aneurysm (i.e., a weakened, thinned-out aneurysm wall), and will not rupture under normal circumstances.

2. *Newtonian characterization of blood*. Blood is an extremely complex fluid and difficult to simulate. This is due primarily to its viscosity. The viscosity of a fluid is a physical property that is dependent on the friction of its component molecules. The shear stress required to overcome this frictional force and move these component molecules in a positive direction is given by

$$S = \eta \frac{dv}{dy},$$

where dv/dy is the velocity gradient. Any fluid that exhibits the linear relationship between the shear stress S and the velocity gradient dv/dy (or dv/dr in cylindrical coordinates) is termed a *Newtonian fluid*. Viscosity is described in terms of the poise, where 1 poise (P) is equal to 1 dyn s/cm^2. In the majority of experiments with aneurysm phantoms, blood is typically modeled as a fluid with a density of 1.056 g/ml and a viscosity of 3.5 cP, representative of a Newtonian fluid. Blood, however, experiences changes in the shear stress through a cross section of an artery or aneurysm and represents a non-Newtonian fluid. A non-Newtonian fluid exhibits a nonlinear relationship between shear stress and velocity gradient at low shear rates since it requires a small initial yield stress to initiate flow. For *in vitro* vascular phantoms, aqueous solutions of polyacrylamide fluids have been used to simulate the non-Newtonian behavior of blood.[38,74,107] Although there are differences in the composition and physiology between human and swine blood, the availability of blood is a distinct advantage in the investigation of experimentally constructed animal models. In addition, mathematical relations similar to those described in Chap. 4 can be implemented into computational simulations to model the viscosity of a non-Newtonian fluid in terms of the shear rate.[108] Although it has never been researched, it is almost certain that a non-Newtonian fluid would have an impact on the resonant frequency due to the increase in shear stress of the aneurysm wall.

3. *A linearly elastic wall*. This factor can be accommodated into all investigative techniques of experimental aneurysms. Biomathematical aneurysm models are typically based on Laplace's law which, in effect, represents a linear relation between stress and strain. However, biomechanical analysis of tissue from aneurysms have been shown to exhibit nonlinear elasticity. The nonlinear behavior of the aneurysm wall tissue characterizes more accurately the aneurysm but would require the solution(s) of a nonlinear, nonhomogeneous differential equation that can be intensive computationally. Similar problems are also encountered in computational simulations of aneurysm hemodynamics. Implementation of a nonlinear aneurysm wall into the equation of motion would also probably increase the natural frequency of the aneurysm, making resonance a much more unlikely phenomenon. A more likely scenario for aneurysm rupture involves the thinning of the aneurysm wall as the radius increases, coupled with structural fatigue due to the forced vibrations from the pulsatile hemodynamic forces. The

availability of a variety of plastic and nonlinearly elastic materials improves the investigational capacity of *in vitro* aneurysm models. Since *in vivo* models generally involve the surgical grafting of venous specimens, the nonlinear behavior of the aneurysm is maintained to some extent. However, this introduces another problem, which will be briefly mentioned here. The surgical grafting of venous tissue to an arterial site is termed an *arteriovenous anastomosis* and introduces a structural change in the mechanical and elastic properties of the surgically altered vessel. The change in intrinsic properties of the vessel manifests itself as *compliance mismatch*, which, in turn, fosters the development of abnormal hemodynamics, turbulence, and the possible onset of vessel rupture.

4. *Collagen and elastin infrastructure*. The microscopic structure of the aneurysm wall is simplified by the incorporation of a single layer of intertwined collagen and elastin networks. Elastin is similar to collagen in that it is also a biomolecule that plays an integral role in the structure and stability of the vessel wall. However, it is considerably weaker than collagen and is believed to be fragmented in the early stages of aneurysm development. The Young's modulus of elastin is 5×10^6 dyn/cm^2. Although the alignment of collagen and elastin in the vascular wall is extremely complicated,[109] this may be an adequate assumption since it is the components of collagen and elastin that account for the elasticity and stability of the aneurysm. A further improvement could be to include a mathematical representation of the active tension produced by smooth muscle cells. Although attempts have been made to model the stress–strain relations of collagen and elastin in terms of physical variables unique to their structural behavior, few studies have incorporated the influence of collagen into mathematical models of aneurysms involving the parametric evaluation of the biophysical and geometric variables.[66,67] In order to model accurately the viscoelasticity of the aneurysm, a mathematical expression is needed that represents stress as a cumulative function of the stress in all previous time frames.[110,111]

6.4   Saccular Aneurysms

The majority of the cerebral aneurysms encountered are berry or saccular aneurysms, which occur most often at points of bifurcation along the large cerebral arteries. Saccular aneurysms develop as a dilatation from a localized region of the vascular wall as opposed to fusiform aneurysms, which are a focal dilatation of the entire region of the vascular wall. Fusiform aneurysms will be discussed later in this chapter. We will now examine how saccular aneurysms might develop, grow, and rupture in terms of both mathematical and physical models.

6.4.1   Initiation of Saccular Aneurysms

At an arterial bifurcation [Fig. 6.6(A)], a parent artery with a radius R_p branches into two daughter arteries with radii R_{d1} and R_{d2} positioned at

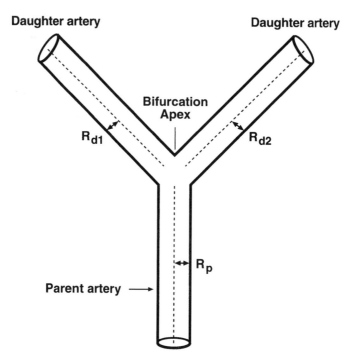

FIGURE 6.6. Arterial bifurcation geometry. (A) Aneurysms occur most often at the apex of an arterial bifurcation, where blood from a parent artery of radius R_p is directed into two daughter arteries of radii R_{d1} and R_{d2}. (B) One reason for the vulnerability of the bifurcation apex is evident in this vector diagram of tensile forces of the parent artery F_p acting down and the duaghter arteries F_{d1} and F_{d2} acting at θ_1 and θ_2, respectively. (Reprinted, with permission, from G.J. Hademenos, The physics of cerebral aneurysms, *Physics Today* **48**, 24–30, 1995.)

angles θ_1 and θ_2, respectively, with respect to the plane bisecting the bifurcation apex. At a normal human arterial bifurcation, blood flow proceeds from the parent artery, is separated at the bifurcation, and is redirected into two daughter arteries of different radii stemming at different angles from the apex. The bifurcation angle of the carotid artery bifurcation ranges from 30° to 120°.[112] The apex of bifurcations is the site of maximum hemodynamic stress in a vascular network because of the impact, deflection, and separation of the blood flow streamlines and vortex formation at the lateral angles.[113] Over a period of years, oscillatory hemodynamic forces, continually striking at the apex of the bifurcation, exert large shear stresses against the endothelial surface and the underlying elastin network, which may cause focal or localized degeneration of the internal elastic lamina and may lead to aneurysm formation.[114] The effect of the hemodynamic forces may be compounded by the presence of abnormal physiological conditions such as chronic high blood pressure (arterial hypertension). The angles and overall geometry of the bifurcation have an effect on the Reynolds number

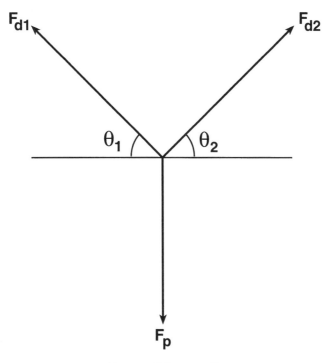

FIGURE 6.6 (*continued*)

as well, increasing the probability of turbulence and the initiation of an aneurysm. The Reynolds number, Re, is a dimensionless hemodynamic parameter used to predict transitions from laminar flow to turbulent flow and is given mathematically as

$$\text{Re} = \frac{v_m d\rho}{\eta},$$

where v_m is the mean blood flow velocity, d is the diameter of the vessel, ρ is the density of blood, and η is the viscosity of blood. Due to a reduction in artery radius and blood flow velocity and an increase in total cross-sectional area, arterial bifurcations lower the critical Reynolds number from 2300 to approximately 400. Thus, turbulent flow is more likely at a bifurcation.

6.4.1.1 Static Equilibrium of an Arterial Bifurcation

In addition to the imposed hemodynamic stress, another factor involved in the increased frequency of aneurysms at a bifurcation apex is the resultant force acting on the apex in static equilibrium. The resultant force exerted on the apex of a bifurcation reduces to a standard problem in mechanics and can be determined easily by resolving the tensile or stretching forces produced by the parent artery, F_p, and the daughter arteries, F_{d1} and F_{d2}. The vector diagram of the forces acting on the apex is given in Fig. 6.6(B). At static equilibrium, it is assumed that no other external forces are acting on the apex and thus the following is true:

$$F_p - F_{d1} - F_{d2} = 0. \tag{6.1}$$

Resolving the three force vectors into their geometrical components, for the x axis,

$$(F_{d1})_x + (F_{d2})_x = 0, \tag{6.2}$$

$$F_{d1} \cos\theta_1 + F_{d2} \cos\theta_2 = 0, \tag{6.3}$$

and for the y axis,

$$(F_{d1})_y + (F_{d2})_y - (F_p)_y = 0, \tag{6.4}$$

$$F_{d1} \sin\theta_1 + F_{d2} \sin\theta_2 - F_p = 0. \tag{6.5}$$

If the magnitude of one of the forces is known, the two other forces can be determined easily by simultaneously solving Eqs. (6.3) and (6.5). It can be seen from Eqs. (6.3) and (6.5) that for a large bifurcation angle (i.e., $\theta_1, \theta_2 \rightarrow 0°$), the forces exerted by the daughter arteries will offset one another exactly with the parent artery responsible for the only force contribution in the y direction. This translates into a larger force exerted on the apex and hence a larger transmural pressure or pressure gradient across the vessel wall, which has been shown to induce shape changes at the apex of arterial bifurcations.[115]

6.4.1.2   Hemodynamics at an Arterial Bifurcation

At a normal human arterial bifurcation, blood flow proceeds from the parent artery, is separated at the bifurcation, and is redirected into two daughter arteries of different radii stemming at different angles from the apical pole as shown in Fig. 6.7. Fully developed flow within the parent artery is represented by a parabolic approximation of the velocity profile given in Chap. 4 as

$$v = \left[\frac{(\Delta P)R^2}{4vL} \right] \left[1 - \left(\frac{r}{R}\right)^2 \right], \tag{6.6}$$

where R is the radius of the vessel and r is the radial position across the cross-sectional radius of the artery. Wall shear stress can be derived from Eq. (6.6) by the following:

$$\tau = -\eta \frac{dv}{dr} = -\frac{(\Delta P)r}{2L}, \tag{6.7}$$

where η is the fluid viscosity. From Eq. (6.7), it can be seen that the wall shear stress increases as the radial position moves from the center to the vessel wall. Thus, at the center of the vessel, wall shear stress is zero and is maximum at the wall. Once blood flow reaches the daughter arteries, notice the increased wall stress exerted on the apex. This is depicted by the asymmetric velocity profiles, skewed toward the apical region. The pressure exerted on the apex is much greater than the pressure exerted on all other sites

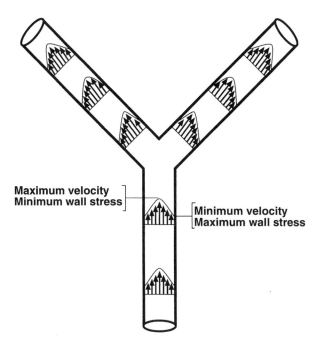

Maximum velocity
Minimum wall stress

Minimum velocity
Maximum wall stress

FIGURE 6.7. Hemodynamic forces play a major role in weakening the apex of an arterial bifurcation. As blood passes from the parent artery into the daughter arteries, which have a combined cross-sectional area greater than that of the parent, it flows more slowly. Excess kinetic energy and momentum of blood leaving the parent artery must be dissipated in the apical region, damaging arterial tissues and promoting turbulent flow of the blood. In this velocity profile, the length of the arrows corresponds to the velocity of blood flow. The higher blood velocities near the apex make this region the site of maximum hemodynamic stress. (Reprinted, with permission, from G.J. Hademenos, The physics of cerebral aneurysms, *Physics Today* **48**, 24–30, 1995.)

of the vessel wall. In addition, experimental studies have shown shape changes at the bifurcation apex in response to changes in the transmural pressure.[115] This pressure difference translates into a higher pressure differential as blood flows from the parent to daughter arteries, increasing the volumetric rate of blood flow in addition to blood pressure.

In order to discuss properly the hemodynamics at an arterial bifurcation, it is important at this stage to understand the behavior of blood flow at points before and after the bifurcation apex.[116] By the conservation of mass, the rate of blood flow, Q, into and out of the bifurcation must be equal or

$$Q_p = nQ_d, \tag{6.8}$$

where Q_p and Q_d represent the flow in the parent and daughter vessels, respectively, and n is the number of branches or daughter vessels in the bifurcation ($n = 2$ for a bifurcation, $n = 3$ for a trifurcation, etc.). The blood flow rate is defined as the mean velocity v times the cross-sectional area of the cylindrical vessel, and hence Q_p and Q_d can be expanded according to

$$Q_p = v_p \pi R_p^2; \quad Q_d = v_d \pi R_d^2. \tag{6.9}$$

Substituting Eq. (6.9) into Eq. (6.8) yields, for a bifurcation,

$$v_p \pi R_p^2 = 2 v_d \pi R_d^2.$$

Rearranging reveals the following:

$$\frac{v_p}{v_d} = \frac{2 R_d^2}{R_p^2} = \text{const.} \tag{6.10}$$

Thus, the mean velocity of blood flow in the daughter vessels will be less than the velocity in the parent vessel by a factor of $(\text{const})/2$.

Another hemodynamic factor integral to the description of blood flow in a bifurcation is the pressure gradient. Blood flow in a vessel can be approximated sufficiently according to Poiseuille's formula:

$$Q = \frac{\pi \Delta P r^4}{8 L \eta},$$

where Q is the flow rate, ΔP is the pressure gradient, r is the inner radius of the vessel, L is the length of the vessel, and η is the blood viscosity. In a manner similar to the blood flow, the pressure gradients in the parent and daughter vessels can be compared using Poiseuille's equation:

$$Q_p = \frac{\pi R_p^4 \Delta P_p}{8 \mu L}; \quad Q_d = \frac{\pi R_d^4 \Delta P_d}{8 \mu L}. \tag{6.11}$$

Substituting Eq. (6.11) into Eq. (6.8) gives

$$\frac{\Delta P_p}{\Delta P_d} = \frac{2 R_d^4}{R_p^4} = (\text{const})^2 / 2. \tag{6.12}$$

In other words, the pressure gradient is greater in the daughter vessels of the bifurcation.

The transfer of hemodynamic energy, momentum, and force produced by increases in the pressure gradient and flow velocity acts to physically degrade the artery wall in the surrounding apical region. This can be visualized through a qualitative analysis of the physical variables involved. The large blood flow velocity in the parent artery translates into a large kinetic energy acting directly on the bifurcation apex. The kinetic energy of an incompressible fluid over a unit volume V is proportional to the flow velocity squared, given mathematically by

$$E_k = \frac{\rho}{2} \int v^2 \, dV. \tag{6.13}$$

Thus, as blood flow from the parent artery enters the daughter arteries, the larger total cross-sectional area of the daughter arteries promotes a decrease in blood flow velocity and a corresponding decrease in kinetic energy as a result of the frictional stress imposed by the blood flow on the artery wall. The dissipation of kinetic energy at the apex of the bifurcation results in structural fatigue of the artery wall and is an important factor in the origin of aneurysms.

The impact of the blood flow on the apex can be better understood by introducing the physical concept of impulse. Impulse is, in effect, the magnitude of an applied variable force which strikes a certain surface area or point over a defined time interval, $t_2 - t_1$, representing one cardiac beat cycle and defined as

$$\mathbf{J} = \int_{t_1}^{t_2} F \, dt, \tag{6.14}$$

where $\mathbf{J}$ is the impulse. The magnitude of the variable force is the ratio of the change in momentum to the change in time. The magnitude of the variable force F is

$$F = \frac{\Delta p}{\Delta t}, \tag{6.15}$$

where Δp is the change in momentum and Δt is the change in time. Thus, from the equation above, it can be reasoned that the impulse becomes large in magnitude if striking a small area (apex) over a small time interval. This equation and overall concept can be used to illustrate the forces generated, not solely by the magnitude of blood pressure but by sudden surges in blood pressure due to physical exertion and daily activities. A sudden increase or surge in blood pressure would increase the magnitude of the variable force [Eq. (6.15)] exerted on the artery wall, which, in turn, increases the magnitude of the impulse. One would run a higher risk of imposing damage to the artery wall by intense exercise as opposed to a steady level of higher blood pressures. Another factor of physical importance is the vibrational energy transferred to the arterial wall by the blood flow.[117,118] Continual vibrations, a result primarily of the forced vibrations or the oscillations that occur in response to the periodic pressure pulse propagated at the onset of a heartbeat, tend to weaken the structural integrity of the bifurcation apex and magnify the existing state of destructive fatigue.

Hemodynamic forces can play a role in the destruction and degradation of the arterial wall in the major cerebral arteries and accelerating the onset and progression of vascular disease. Arterial regions that are exposed to large blood flow velocities over an extended period of time are sites of high probability for irreversible damage to the histological structure of the vascular wall. A widely accepted contributing factor in each stage of aneurysm growth is based primarily on hemodynamics and its influence on resulting biophysical phenomena and interactions[39,41] that occur as a direct result of blood flow within irregular arterial geometries such as tortuous segments, bends, and bifurcations commonly encountered in the human cerebral circulation. These interactions involve the static tensile forces exerted on vulnerable points of the artery geometry, the viscous forces, pressure forces, velocity, kinetic energy, momentum, and impulse presented by the bloodstream, and the shear stress and vibrational displacements experienced by the elastic artery wall in response to distending forces. The structural conditions presented by such geometries coupled with the effects of the biophysical variables are conducive to abnormal flow behavior and patterns resulting in the degradation of the artery wall and the formation of a bulge or small pouch at the damaged region of the artery wall.

The exact influence of each of these hemodynamic factors on the initiation and development of an aneurysm is unknown. From a biomechanical standpoint, the resultant forces from these hemodynamic factors must be sufficient to alter irreversibly the elastin component of the arterial wall.[119,120] Once the bulge has been initiated, a pocket of highly disturbed secondary flow, referred to in the literature as turbulence,[121,122] develops first within the branches of the bifurcation apex and continuing within the dilatation. Once the arterial wall defect has been introduced, growth becomes imminent and is related directly to secondary physiological and health factors such as hypertension,[30,31] the presence of other aneurysms and other vascular diseases, i.e., a factor that could adversely affect the balance of intravascular pressures distributed within the local vessels. As the aneurysm develops and grows, it displaces the tethered tissue and increases the extravascular pressure exerted on the aneurysm wall. The magnitude of this increase in pressure is dependent on the surrounding tissue and can be determined only by experimentation with the tissue in question.

6.4.2 Development of Saccular Aneurysms

A saccular aneurysm may be considered conveniently as a spherical dilatation of the arterial wall that develops at the apex of a bifurcation. The geometrical parameters used to describe an aneurysm are shown in Fig. 6.8. In the figure, the structural components and forces of the aneurysm are depicted, which include a thin-walled spherical dome or fundus of radius R and a uniform wall thickness h with a pressure P exerted against the aneurysm wall. In terms of the pressures observed during the cardiac cycle, researchers have shown that the intra-aneurysmal pressure correlates well with the systolic blood pressure.[123-125] Based on physical reasoning, this appears logical since it is during cardiac systole that the blood is pumped at its maximum force. In static equilibrium, the distribution of forces acting on the spherical aneurysm can be explained by Laplace's law.

Laplace's law states that there is a linear relation between the circumferential stress and the radius of an elastic sphere, i.e., the stress required to maintain static equilibrium increases as the radius increases. Laplace's law,

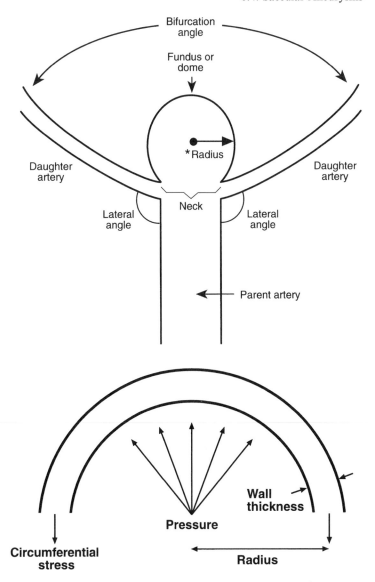

FIGURE 6.8. A cerebral aneurysm develops (A) as a spherical bulge of radius R and wall thickness h. As the arterial tissue is stretched from its original circumferential length L, its elastin fibers break or fragment, leaving collagen to bear the structural load. At equilibrium, the tension within the aneurysm wall opposes the pressure forces (B). Other cerebrovascular disease or hypertension can upset this equilibrium, causing the aneurysm to grow and ultimately rupture. (Reprinted, with permission, from G.J. Hademenos, The physics of cerebral aneurysms, *Physics Today* **48**, 24–30, 1995.)

in effect, equates a radial force P produced by the transmural blood pressure over the cross-sectional area of the sphere to a circumferential stress S directed along the aneurysm wall in the opposite direction that compensates for the distension and is required to maintain equilibrium. In mathematical form, the circumferential stress exhibited by the walls of an elastic sphere is given by Laplace's law:

$$S = \frac{PR}{2h},\qquad(6.16)$$

where R is the radius of curvature and h is the wall thickness. Equation (6.16) is only valid, however, for thin-walled aneurysms, i.e., $h \ll R$. The integral factors that influence the circumferential stress can be shown by taking the differential of Eq. (6.16):

$$dS = d\left(\frac{PR}{2h}\right) = \frac{1}{2}\left(\frac{R}{h}dP + \frac{P}{h}dR - \frac{PR}{h^2}dh\right).\qquad(6.17)$$

From Eq. (6.17), it can be seen that an increase in the stress is dependent not only on an increase in pressure but also on increases in radius and wall thickness as well.

The circumferential stress for an elastic object, given in Eq. (6.16), can also be expressed by Hooke's law of elasticity as

$$S = \varepsilon E,\qquad(6.18)$$

where ε is the circumferential strain and E is the elastic modulus of the aneurysm under physiologic conditions. As stated previously, the circumferential strain exhibited by the aneurysm wall can be expressed as the ratio $\Delta R/R$, where ΔR is the change in radial length between the relaxed and strained states and R is the radial length in the unstrained state. The radial length in the unstrained state is denoted by L in Fig. 6.8. Assuming that changes in the radial length are infinitesimal, Eq. (6.18) can be expressed in differential form as

$$dS = \frac{dR}{R}E.\qquad(6.19)$$

Equating Eqs. (6.17) and (6.19),

$$\frac{1}{2}\left(\frac{R}{h}dP + \frac{P}{h}dR - \frac{PR}{h^2}dh\right) = \frac{dR}{R}E.\qquad(6.20)$$

The growth of the aneurysm can be obtained by determining the differential relation between the radius and pressure, dR/dP. Dividing both sides of Eq. (6.20) by dP,

$$\frac{1}{2}\left(\frac{R}{h}\frac{dP}{dP} + \frac{P}{h}\frac{dR}{dP} - \frac{PR}{h^2}\frac{dh}{dP}\right) = \frac{dR}{dP}\frac{E}{R}.\qquad(6.21)$$

Simplifying and solving for dR/dP,

$$\frac{dR}{dP} = \frac{\left(\dfrac{R}{h} - \dfrac{PR}{h^2}\dfrac{dh}{dP}\right)}{\left(\dfrac{2E}{R} - \dfrac{P}{h}\right)}. \tag{6.22}$$

The growth of the aneurysm can be better visualized qualitatively by considering the increase in aneurysm volume with respect to pressure or the volume distensibility (dV/dP). The volume distensibility can easily be derived from Eq. (6.22) by the following:

$$\frac{dV}{dP} = \frac{dV}{dR}\frac{dR}{dP}, \tag{6.23}$$

where V is the volume of a sphere $(\frac{4}{3}\pi R^3)$. Thus

$$\frac{dV}{dP} = 4\pi R^2 \frac{dR}{dP} \tag{6.24}$$

or

$$\frac{dV}{dP} = 4\pi R^2 \frac{\left(\dfrac{R}{h} - \dfrac{PR}{h^2}\dfrac{dh}{dP}\right)}{\left(\dfrac{2E}{R} - \dfrac{P}{h}\right)}. \tag{6.25}$$

The volume distensibility of the aneurysm or rate of aneurysm volume expansion allows one to investigate accurately the biophysical response of the elastic aneurysm to distending pressures. It can be seen that there is a cubic dependence on the radius and a definite influence from the thickness that decreases as the pressure and hence the radius expands. It is also possible to algebraically manipulate Eq. (6.25) and derive other expressions relating the volume, pressure, thickness, elastic modulus, and radius. These equations can then serve as the basis for the quantitation of experimental observations and biophysical phenomena. It has been mentioned previously that, as a direct result of increasing internal pressure, the aneurysm wall in advanced developmental stages prior to rupture consists primarily of collagen with the elastin fiber contribution fragmented upon initiation and early development. As described previously, Laplace's law describes a linear relation between the circumferential stress and the radius of any elastic curved surface, i.e., the stress required to maintain static equilibrium increases as the radius increases. A particular example of Laplace's law is the balloon, and it thus becomes a logical deduction to draw an analogy and extend this concept to a spherical aneurysm. However, a major difference between the balloon and the aneurysm is the elasticity of the wall material. The balloon is linearly elastic while the aneurysm wall is viscoelastic. Viscoelasticity implies a nonlinear relation between increases in the radius and stress. In other

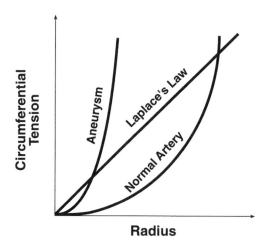

Radius

FIGURE 6.9. Graphical representation of the influence of radius on circumferential tension for tissue obeying Laplace's law, tissue of a normal artery, and tissue of an aneurysm. (Source: G.J. Hademenos, T.F. Massoud, D.J. Valentino, G.R. Duckwiler, and F. Viñuela. A mathematical model for the development and rupture of intracranial saccular aneurysms based on a biomathematical model. *Neurological Research* **16**, 376–384, 1994. Reproduced by permission of the publishers Forefront Publishing.)

words, for the same radial expansion, the aneurysm wall will exert a much larger stress than the balloon. Laplace's law, given by Eq. (6.16), does not account for viscoelasticity of the arterial wall, and thus does not accurately represent a saccular aneurysm. Even for a normal artery, it has been shown that the circumferential stress varies nonlinearly with the radius.[126] Figure 6.9 shows typical curves of the circumferential stress versus radius for (A) an artery that behaves according to Laplace's law, (B) a normal artery, and (C) aneurysm tissue. It should be noted that the curves in Fig. 6.9 assume static properties of stress when, in reality, the dynamic properties of stress are the most important due to the changing mechanical properties of the artery and are the result of its viscoelasticity.

6.4.2.1.   An Equation of Motion for a Saccular Aneurysm

The intracranial aneurysm can be modeled as a thin elastic spherical shell with a radius R and a wall thickness h as shown in Fig. 6.10. Although clinical and histopathologic tissue examination of aneurysms reveal an irregularly shaped aneurysm wall,[48] the aneurysm wall in this study is assumed to be homogeneous, isotropic, and linearly elastic as a first order approximation. The angle ϕ represents the angle of the aneurysm wall closure and N is the aneurysm neck diameter or size of the circular opening within the parent vessel leading to the aneurysm fundus.[61] The aneurysm neck

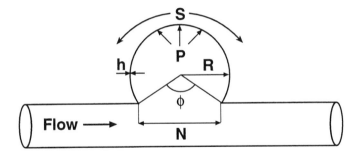

FIGURE 6.10. A cross-sectional view of the saccular aneurysm employed in the bio-mathematical model. P represents intra-aneurysmal pressure; S, circumferential stress; R, aneurysm radius; N, neck diameter; h, wall thickness; ϕ, neck angle (angle of wall closure).

diameter N ranges in size from 1.5 to 10.0 mm and is classified as small ($N \leq 4.0$ mm) or wide ($N > 4.0$ mm) with an average neck diameter $N = 4.0$ mm.[127] The variables ϕ and N are related by the following relation[70]:

$$\frac{N}{2} = R \sin \frac{\phi}{2}. \tag{6.26}$$

Defining the parameter ξ, which is dependent on time, as the displacement at some later time from R to $R + \xi(t)$, the strain ε of the expanded aneurysm is then given by[128]

$$\varepsilon = \frac{S}{E} = \frac{\xi A}{\Sigma R}, \tag{6.27}$$

where A is the area of the aneurysm neck, Σ is the surface area of the aneurysm fundus or dome, and R is the original aneurysm radius. S can also be expressed, according to Laplace's law for a spherical elastic object, as

$$S = \frac{PR}{2h}, \tag{6.16}$$

where P is the intra-aneurysmal pressure.

An equation of motion describing the vibrational displacement of the aneurysm wall can be derived by considering the forces acting on the aneurysm in static equilibrium. These include forces due to pressure ($\mathbf{F}_{pres}$), circumferential stress ($\mathbf{F}_{stress}$), and inertia ($\mathbf{F}_{iner}$). $\mathbf{F}_{pres}$ is a distending or pushing force that acts in an outward direction while the other forces, i.e., $\mathbf{F}_{iner}$ and $\mathbf{F}_{stress}$, are compressive forces and act in a direction opposite to $\mathbf{F}_{pres}$. The balance of forces written in equation form are

$$\mathbf{F}_{pres} = \mathbf{F}_{stress} + \mathbf{F}_{iner}. \tag{6.28}$$

Since the motion of the aneurysm is driven by a pulsatile pressure function, a more accurate characterization of the dynamics of the aneurysm can be obtained by expressing Eq. (6.28) as the balance of pressures (force per unit surface area within the aneurysm) instead of forces:

$$P = P_{stress} + P_{iner}. \tag{6.29}$$

P represents the pulsatile driving force exerted on the aneurysm wall, which is related directly to the intra-aneurysmal pressure.

P_{stress} is the pressure or force exerted on the aneurysm wall by the circumferential stress and can be determined by setting Eq. (6.27) equal to Eq. (6.15) and solving for P:

$$\frac{PR}{2h} = E \frac{\xi A}{\Sigma R}, \tag{6.30}$$

$$P_{stress} = 2hE \frac{\xi A}{\Sigma R^2}. \tag{6.31}$$

Equation (6.31) can be further simplified by

$$P_{stress} = B\xi, \tag{6.32}$$

where $B = 2hEA/(\Sigma R^2)$.

F_{iner} is the inertial force due to the acceleration of the aneurysm wall and is equal to, according to Newton's second law,

$$F_{iner} = ma \tag{6.33}$$

where m is the object mass and a is the acceleration of the object. Since the intracranial aneurysm is blood-filled, the inertial force term consists of the mass of the blood in the aneurysm and mass of the aneurysm wall and is represented by the system's kinetic energy, which is concentrated in the vicinity of the aneurysm neck,[128]

$$F_{iner} = m_b a = \frac{1}{2} m_b \frac{d^2 \xi}{dt^2}, \tag{6.34}$$

where m_b, the mass of blood, can be approximated by

$$m_b \approx 0.8 \rho_b a A \tag{6.35}$$

where ρ_b is the density of blood, a is the aneurysm neck radius, and A is the aneurysm neck area. Also from Eq. (6.34), $d^2\xi/dt^2$ is the acceleration of the blood-filled aneurysm wall in response to an external driving force. The pressure due to inertia, P_{iner}, is

$$P_{iner} = \frac{F_{iner}}{A} = \frac{1}{2}(0.8\rho_b a) \frac{d^2\xi}{dt^2}. \tag{6.36}$$

Similar to the other components of pressure, P_{iner} can be simplified according to

$$\mathbf{P}_{\text{iner}} = A \frac{d^2\xi}{dt^2} \tag{6.37}$$

where $A = \frac{1}{2}(0.8\rho_b a)$.

Substituting the components of pressure given by Eqs. (6.31), and (6.36), into Eq. (6.29) yields the following differential equation:

$$P = \frac{1}{2}(0.8\rho_b a)\frac{d^2\xi}{dt^2} + 2hE\frac{\xi A}{\Sigma R^2}. \tag{6.38}$$

Equation (6.38), which is a linear, nonhomogeneous, second-order differential equation, can be rewritten as

$$P = A\frac{d^2\xi}{dt^2} + B\xi. \tag{6.39}$$

The pressure P can be represented as the driving force of the pulsatile blood flow and is stated mathematically as

$$P = F_{\text{bp}}\cos(\omega_{df}t), \tag{6.40}$$

where F_{bp} is the pulsatile hemodynamic force produced by the systolic blood pressure and ω_{df} is the frequency of the driving force. Equation (6.38) is then rewritten as

$$A\frac{d^2\xi}{dt^2} + B\xi = F_{\text{bp}}\cos(\omega_{df}t). \tag{6.41}$$

Equation (6.41) demonstrates that it is possible to characterize mathematically the vibrational properties of the aneurysm. Solution of the equation, given later in the chapter, yields the spatial location of the aneurysm wall in response to a driving force with respect to time. In addition, the equation can be used to determine the resonant frequency of the aneurysm wall and investigate the effects of external forces on the resonant frequency, providing a physical basis for the influence of resonance in aneurysm rupture. This will also be discussed later in the chapter.

6.4.2.2 Intra-Aneurysmal Hemodynamics

Up to this point, the description of the development of the aneurysm has been limited to the statics and dynamics of the aneurysm wall, neglecting intra-aneurysmal hemodynamics. Blood flow in most saccular aneurysms is regular and predictable according primarily to the geometrical relationship between the aneurysm and its parent artery.[129] As blood flows within the parent artery with an aneurysm, divergence of blood flow, which occurs at the inlet of the aneurysm, leads to dynamic disturbances with a Bernoulli effect producing increased lateral pressure and retrograde vortices that are

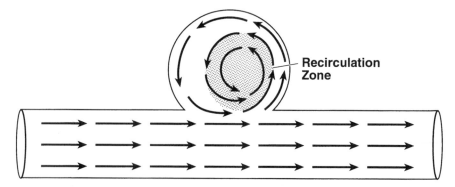

FIGURE 6.11. Schematic diagram showing the hemodynamic circulation patterns within an aneurysm.

easily converted to turbulence.[130] Blood flow proceeds from the parent vessel into the aneurysm at the distal or downstream extent of the aneurysm neck, circulates around the periphery along the aneurysm wall from the neck to the top of the fundus (downstream to upstream), returning in a type of "isotropic shower" along the aneurysm wall toward the neck region, and exits the proximal or closest extent of the aneurysm neck into the parent vessel (Fig. 6.11).[131] As flow persists, areas of stagnation or vortices develops within a central zone of the aneurysm. These rotating vortices, formed at the entrance to the aneurysm at each systole and then circulated around the aneurysm, are caused by the slipstreams or regions of recirculating flow rolling upon themselves when entering the aneurysm at its downstream wall during systole.[132] The stagnant vortex zone occurs in the center and at the fundus or upper portion of the aneurysm and becomes more pronounced in larger aneurysms. These hemodynamic patterns are depicted in Fig. 6.12. It is this stagnant zone that is believed to promote the formation of thrombi or blood clots, particularly in giant aneurysms.

6.4.3   Rupture of Saccular Aneurysms

Rupture of intracranial aneurysms occurs in 28 000 North Americans each year with approximately 50% of these people dying within the first 30 days following rupture.[133] The rupture of a cerebral aneurysm occurs when the tension of the aneurysm wall exceeds the force produced by the structural components. When an intracranial saccular aneurysm ruptures, blood is forced under a pressure of approximately 100 mm Hg into the subarachnoid space, where a pressure of 0 to 10 mm Hg is prevalent under normal circumstances.[134] The aneurysm wall, fully distended prior to rupture, consists primarily of collagen, which bears the majority of the mechanical response against the stress induced by the tension. The strength of the aneurysm wall